Basic Values
on Single Span Beams

Tables for calculating continuous beams and frame
constructions including prestressed beams

By

Günter Baum

Dipl.-Bauing. E. T. H. / S. I. A. Zürich

With 59 figures

Springer-Verlag

Berlin / Heidelberg / New York

1966

A German edition of this book has been published in 1965 entitled:

Grundwerte am Einfeldbalken
von Günter Baum

ISBN-13: 978-3-642-49163-4   e-ISBN-13: 978-3-642-49161-0
DOI: 10.1007/ 978-3-642-49161-0

Library of Congress Catalog Card Number 65–27 225

Title No. 1326

Preface

In keeping with the general trend towards rationalisation, static calculations have of late also been programmed by electronic computers. The number of problems which can be advantageously resolved in this way is, however, very limited as yet, partly on account of the relatively high cost involved and partly due to the waiting time the statician must suffer after collecting together his data and, finally, because the programming possibilities of the computer are limited.

Nonetheless, if static calculations have to be rationalised, there is another way: all beam structures—whether they be continuous beams or frame constructions—are arithmetically based on individual spans which are freely supported or fixed at the ends. If the basic values for these can be ascertained quickly and accurately, then a considerable part of the arithmetical work is already done. It is the aim of this work to provide the statician with these values. An attempt has been made to deal as comprehensively as possible with all the cases of loading likely to arise in practice. Naturally, one case or another is bound to happen more frequently whilst others are seldom encountered. However, this allembracing programme is intended to make it possible for the user of this work, after a brief, familiarising period, always to use the same arithmetical procedure, the choice of the actual method being left to him.

The present work is the result of the continual extension of tables which I compiled for my own engineering office; it has been compiled from practical experience and is designed for practising staticians.

At this point, I wish to express my thanks to Herr Dipl.-Ing. H. DALSHEIM for the valuable help he has given me during the treatment of basic values in the case of prestressing of concrete.

Special thanks are due to my publishers for the extremely expedient and attractive way they have presented this book.

Zurich, December 1965

Günter Baum

Contents

No.	Loading	No.	Loading	No.	Loading
1		15		28	
2		16	Quadr. parabola	29	
3		17	Quadr. parabola	30	M
4		18	Quadr. parabola	31	M M M M
5		19	Quadr. parabola	32	M M M M
6		20	Quadr. parabola	33	δ_1 δ_2
7		21	Quadr. parabola		δ_1 δ_2
8		22		34	t_0 t_u
9		23		35 Whole Span Loadings	
10				V1	Quadr. parabola
11		24		V2	
12		25		V3	Cubic parabola
13		26	Quadr. parabola	V4	
14		27	P	V5	

Introduction

A. General remarks

It is the intention of this work to provide practising staticians with already-calculated basic values for beams which are simply supported or which are fixed at one or both ends, taking into account virtually all the cases of loading possible in practice.

By using these basic values, it becomes a simple matter to treat continuous beams, subject to any form of loading, by the moment distribution method (CROSS, KANI etc.) or the three-moment theorem (CLAPEYRON).

The following values are considered to be *basic values*:

1. Reactions A_0 and B_0 at the two simple supports of a singlespan beam.

2. The moments of the simply-supported beam

M_{0a} at the starting point of a line load
M_{0s} the terminal point of a line load
$M_{0\,\mathrm{max}}$ the maximum moment.

3. The abscissa x_m, where the sharing forces Q_0 become nil or maximum beam moment $M_{0\,\mathrm{max}}$ occurs.

4. The angles α_0 and β_0 of rotation of the left or right end respectively of the simply-supported beam, multiplied by EI.

5. The fixed-end moments (FEM's) of the beam, fixed at one or both ends:

M_1 the left FEM of a beam fixed at both ends
M_2 the right FEM of a beam fixed at both ends
M_1^0 the FEM of a beam with one end fixed, the other pinned, a second index, in the case of a line load, indicating whether the line load starts at the built-in end (M_{1A}^0) or at the free support (M_{1B}^0).

The cases of loading dealt with start from a general approach. They have been arranged from the point of view of the special connection of the load diagram to particular points (starting, terminal and symmetry point) of the beam, first the *bound* and then the entirely *free* loadings and in conclusion the cases of prestress being given.

In order to ensure that the coefficient values are represented in tabular form with sufficient clarity, only those loadings can in principle be considered which can be represented by two parameters. For example, a trapezoidal line load of variable load intensity (q) and a free commencement of the line load can only be shown by three parameters. In this case, it is simpler to superimpose the values of a uniformly distributed load and those of a triangular load, than to give a multiplicity of tables which would undermine the handiness of the work.

Recently, prestressed concrete has gained considerably in importance. If we disregard the peculiarities in designing prestressed cross-sections, prestressed concrete entails, in beam statics, the essential new component of determining the moments which prestressing creates at the supports. In the various countries and literature, these moments are variously designated: *Statisch unbestimmte Vorspannmomente*[1] or *Umlagerungsmomente*[2] or *Parasitäre Momente*[3], all of which terms are best rendered in English as *secondary moments*. It is entirely desirable for purposes of practical designing, to be able to estimate these moments, which is why these values and also the values which are dependent upon them have been considered as *basic value* and are shown in tabular form for those cases such as arise with continuous beams, pre-supposing that a prestressing tendon is composed of various parts, these parts in every case intersecting with a horizontal tangent.

The cases of rectilinear prestressing tendons are so simple that tables are unnecessary. In practice, for the simply-supported ends of continuous beams, prestressing tendons according to a quadratic parabola are mainly used, whilst the prestressing tendons of intermediate supports comply with cubic parabolae.

Friction losses on prestressing force have been basically omitted, not only by reason of the considerable mathematical complications, but particularly on account of the friction laws which have to be taken into account according to the specifications or make of the prestressing tendons and which would have furnished a third parameter which cannot be shown in a single, easily-understood table.

The *basic values* which relate to the secondary moments are:

1. The FEM's $\overline{M}_1$ and $\overline{M}_2$ over the left and right supports of a single-span beam fixed at both ends.

2. The FEM $\overline{M}_2^0$ of a beam built-in at one end (right).

3. The shearing force $\overline{Q}$ created by $\overline{M}_1$ and $\overline{M}_2$.

4. The shearing force $\overline{Q}_0$ created by $\overline{M}_2^0$.

5. The angles $\bar{\alpha}_0$ (left) an $\bar{\beta}_0$ (right), multiplied by EI, occurring on a simply-supported beam with both ends pinned.

6. The abscissa $\bar{x}_0$ of the intersection between the axis of the beam and the prestressing tendon, laid according to a cubic parabola (eccentricity $e_x = 0$).

The complete calculation of tabular values is intended to take a large part of the practising statician's routine work off his shoulders and eliminate from actual calculation work as many errors as possible which can very easily occur in dealing with an individual case as compared with calculating tables. Moreover, the present tables will avoid any simplified load assumptions made due to the general complexity of handling effective loadings and which do not conform to reality.

The tables relate to a twenty-fold graduation of the span. Each table is preceded by a corresponding diagram and the theoretically accurate formulae. The structure of the formulae was determined by the following considerations:

a) By reason of the essential general validity, the lengts are of necessity given by ratios related to the span width.

b) Units of loading have been left in their original form, as is required, due to the differing physical dimensions and the desirability of applying various methods of calculation.

c) Basically, the tabular values represent non-dimensional coefficients.

d) All the indices are always used only for determining one single value, to eliminate mistakes.

e) In principle, the following rules apply to the sign:

1. Clockwise moments are positive.

2. Shearing forces are positive if directed upwards to the left of the intersection.

3. Angles of rotation are positive if the deformed central axis of the beam is below the undeformed axis.

---

[1] LEONHARD, F.: Spannbeton für die Praxis, Berlin: Ernst & Sohn 1955.
[2] HAHN, J.: Spannbeton, Theorie und Bemessung, Düsseldorf: Werner-Verlag 1960.
[3] This is the term generally used in Switzerland.

All values relate to a moment of inertia which is constant for every span, which is predominantly the case in present-day building.

Since the coefficients are mainly < 1, no zero appears before the decimal point, for greater simplicity.

In consideration of the great importance of reinforced concrete in modern building, a table is included at the end which gives the ratios of moments of inertia of the T-cross-section beam to those of a rectangular cross-section beam of the same dimensions, without the slab. This representation seems to be the most favourable for the various methods which can be employed, as desired.

For cases where the formulae are applied directly, a table is included giving the various powers of n, because these values predominate.

In the introduction to the book, along with a few notes on the reciprocal dependency of the basic values and a review of the differential geometric relations of static values in the cross-section and the functions of deformation, a series of specimen applications are given, intended to facilitate use of the tables by the beginner. Practitioners, too, would do well to read the contents before using the tables.

When the tabular values were determined, the moments at the fixed ends were ascertained on a basis of the relationships set out in Section B, the angles of rotation of the ends (terms due to the load) being calculated by means of the equation of virtual work $\alpha_0 = \int \frac{M\,M'}{E\,I}\,\mathrm{d}\,x$.

Where beam constructions are concerned, the effect of the shearing forces is unimportant. For movable frame constructions, where the influence of shearing forces must not be disregarded, due allowance must be made for the particular method applied.

For comprehension of the signs used, the reader is referred basically to the illustrations and formulae which precede the appropriate tables.

In all cases, the angles of rotation are the values multiplied by $E\,I$.

B. Relationships between fixed end moments and deformation values in the simply-supported beam

1. The bilaterally fixed span under transversal loads

For the young statician who knows CLAPEYRON's theorem of three moments for continuous beams, Fig. 1, where, for constant moment of inertia, it takes the form

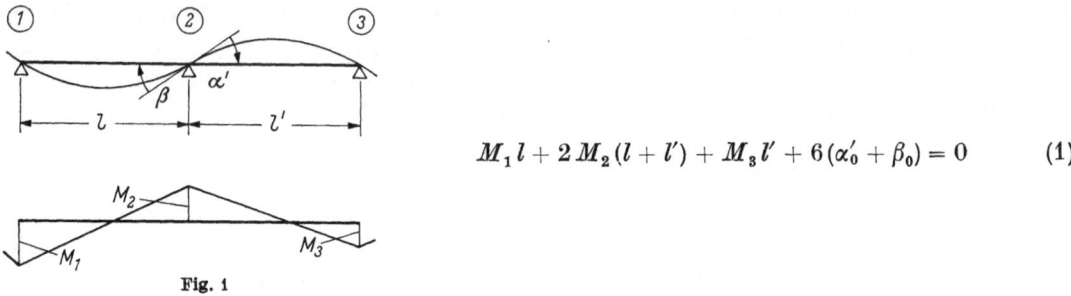

$$M_1\,l + 2\,M_2\,(l + l') + M_3\,l' + 6\,(\alpha'_0 + \beta_0) = 0 \qquad (1)$$

Fig. 1

it often happens that using the equation for this case presents difficulties in connection with imagination, due to the disappearance of the angles at the support when the ends are fully built-in. This can be overcome by the following artifice:

If the middle span $\overline{BC}$ of length l_2 of the three-span beam shown in Fig. 2 is examined, it will immediately be realised that the angles of rotation α and β are smaller than would be the case with a single beam BC with no lateral spans. If we assume for the time being that only the span BC is loaded, which would be the case anyway if there were no lateral spans (Fig. 2 b) for the fixed span l_2, it is immediately obvious that the lateral spans l_1 and l_3 act as elastic fixings on the middle span l_2. The degree of elasticity of these fixings is determined by the beam properties of the lateral spans.

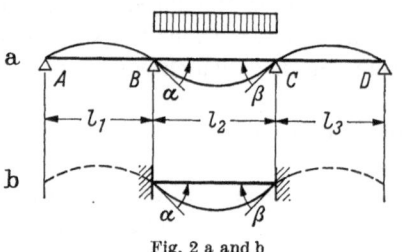

Fig. 2 a and b

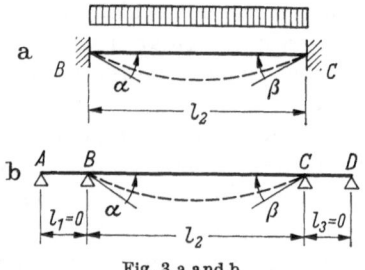

Fig. 3 a and b

A beam is more elastic (or *softer*), the longer it is, the smaller its moment of inertia and the smaller its modulus of elasticity. The elasticity of the beam is therefore directly proportional to the quotient $\dfrac{l}{EI}$. Since the elasticity of the beam and the elasticity of the fixing of an adjacent span as represented by this beam are identical, we can say:

a) If $\dfrac{l}{EI} = 0$, then the elasticity of the fixing is nil, i.e. we have a **rigid** or **absolute fixing**. Since $0 < \dfrac{l}{EI} < \infty$, $\dfrac{l}{EI}$ can only be 0 for $l = 0$.

In other words: a rigid (or absolute) fixing can be replaced by an adjacent span where $l = 0$.

b) If $\dfrac{l}{EI} = \infty$, then there is unlimited elasticity in the fixing. This is the case for $l = \infty$.

In other words: a pinned support can be replaced by an adjacent span where $l = \infty$.

Since, in the present work, we are always concerned with absolute fixings, we are representing these here by adjacent or joining spans of the length $l = 0$, as shown in Fig. 3. The substituting spans have no real magnitude of span, and consquently no loads too. Since the points A and B or C and D, as the case may be, can be infinitely close to each other, then in this case the tangent on the deflection curve is identical with the axis through the beam, i.e. it is horizontal, in other words it satisfies the conditions of rigid fixing. For $l = 0$, the value EI only has to satisfy the condition of having a finite magnitude for $\dfrac{l}{EI}$ to equal 0, so that EI of the fixing spans can in particular also have the same value as the fixed span itself. Therefore, for our purposes, the special form of CLAPEYRON's theorem three moments can readily be applied for a constant moment of inertia. If we do this for point B, then it follows:

$$M_A l_1 + 2 M_B (l_1 + l_2) + M_C l_2 + 6 (\beta_{01} + \alpha_{02}) = 0 \tag{2}$$

and for point C:

$$M_B l_2 + 2 M_C (l_2 + l_3) + M_D l_3 + 6 (\beta_{02} + \alpha_{03}) = 0 . \tag{3}$$

In which

$$M_A = 0 \qquad\qquad l_3 = 0 \qquad\qquad \beta_{01} = 0$$
$$l_1 = 0 \qquad\qquad M_D = 0 \qquad\qquad \alpha_{03} = 0 .$$

Which gives us:

$$2 M_B l_2 + M_C l_2 = -6 \alpha_{02} \tag{2'}$$

$$M_B l_2 + 2 M_C l_2 = -6 \beta_{02} \tag{3'}$$

or

$$\alpha_{02} = - \frac{(2\,M_B + M_C)\,l_2}{6} \tag{2''}$$

$$\beta_{02} = - \frac{(M_B + 2\,M_C)\,l_2}{6}\,. \tag{3''}$$

Since the indices "2" used always relate to the only real span, we can omit them and write:

$$\alpha_0 = - \frac{(2\,M_B + M_C)\,l}{6} \tag{4}$$

$$\beta_0 = - \frac{(M_B + 2\,M_C)\,l}{6}\,. \tag{5}$$

For cases where we have a beam with only one built-in end whilst the other is simply supported, in Fig. 3 for example at B or C, the corresponding moments at the fixed ends M_B or M_C are equal to zero and we obtain:

with an articulated support at C, for $M_C = 0$ and

$$\alpha_0 = - \frac{1}{3}\,M_B^0\,l \tag{6}$$

or, in the event of the pinned support being at B and $M_B = 0$:

$$\beta_0 = - \frac{1}{3}\,M_C^0\,l. \tag{7}$$

All the cases of loading must abide by the general laws expressed in equations (4) to (7).

From equations (4) to (7), by simple interchange of the appropriate values, it is possible to obtain the following relations between the fixed end moments and the angles of rotation which, taking into account the representation with the coefficients in the tables, become analogous relationships among the latter:

We have defined:

$$M_1 = -k_1\,q\,l^2 \qquad M_2 = -k_2\,q\,l^2 \qquad M_1^0 = M_{1A}^0 = -k_1^0\,q\,l^2 = -k_{1A}^0\,q\,l^2$$

$$\alpha_0 = k_9\,q\,l^3 \qquad \beta_0 = k_{10}\,q\,l^3 \qquad M_2^0 = M_{1B}^0 = -k_{1B}^0\,q\,l^2.$$

Since only the absolute amount of the coefficients k interests, thus k always being considered as positive, then whatever the nature of the loading, the following equations will generally apply:

$$\alpha_0 = - \frac{2\,M_1 + M_2}{6}\,l \qquad\qquad k_9 = \frac{2\,k_1 + k_2}{6}$$

$$\beta_0 = - \frac{2\,M_2 + M_1}{6}\,l \qquad\qquad k_{10} = \frac{2\,k_2 + k_1}{6}$$

$$\alpha_0^0 = - \frac{1}{3}\,M_1^0\,l \qquad\qquad k_9 = \frac{k_1^0}{3} \quad \text{resp.} \quad \frac{k_{1A}^0}{3}$$

$$\beta_0^0 = - \frac{1}{3}\,M_2^0\,l \qquad\qquad k_{10} = \frac{k_{1B}^0}{3}$$

$$M_1 = - \frac{1}{l}\,2\,(2\,\alpha_0 - \beta_0) \qquad\qquad k_1 = 2\,(2\,k_9 - k_{10})$$

$$M_2 = - \frac{1}{l}\,2\,(2\,\beta_0 - \alpha_0) \qquad\qquad k_2 = 2\,(2\,k_{10} - k_9)$$

$$M_1^0 = M_{1A}^0 = - \frac{1}{l}\,3\,\alpha_0 \qquad\qquad k_1^0 = k_{1A}^0 = -\,3\,k_9$$

$$M_2^0 = M_{1B}^0 = -\frac{1}{l}\,3\,\beta_0 \qquad\qquad k_{1B}^0 = -\,3\,k_{10}$$

$$M_1^0 = \frac{2\,M_1 + M_2}{2} = M_{1A}^0 \qquad\qquad k_1^0 = k_{1A}^0 = \frac{2\,k_1 + k_2}{2}$$

$$M_2^0 = \frac{2\,M_2 + M_1}{2} = M_{1B}^0 \qquad\qquad k_{1B}^0 = \frac{2\,k_2 + k_1}{2}$$

$$M_1 = \frac{1}{2}\,(2\,M_1^0 - M_2) \qquad\qquad k_1 = \frac{1}{2}\,(2\,k_1^0 - k_2)$$

$$M_2 = 2\,(M_1^0 - M_1) \qquad\qquad k_2 = 2\,(k_1^0 - k_1)$$

As can be readily appreciated from the above expressions, any value in one column can be obtained directly two values in the particular column are known. This corresponds to the second degree of indetermination of a bilaterally fixed beam under vertical loads ($H = 0$).

2. Settling of supports of a beam with both ends built-in

In order to allow for the settling of supports, we imagine the beam to be unloaded, but the supports on the other hand having settled, as shown in Fig. 4. Pre-supposing small amounts of support settlements (tg $\varphi \approx \varphi$), then, along the same line as Eq. (1), CLAPEYRON's equation for $E\,I$ constant is:

$$M_1\,l + 2\,M_2\,(l + l') + M_3\,l' + 6\,E\,I\left[\frac{\delta_1 - \delta_2}{l} + \frac{\delta_3 - \delta_2}{l'}\right] = 0 \qquad (8)$$

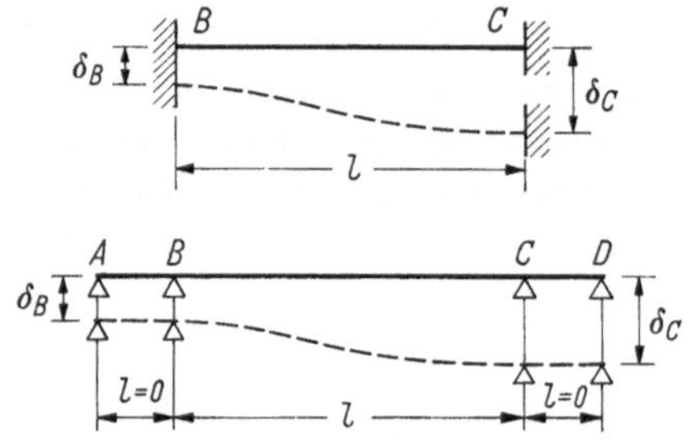

Fig. 4

or, for the case of the beam which is built in at both ends, according to Fig. 4, where the joining spans *replacing* the fixings must in toto be assumed as having settled to a parallel extent:

Related to B:

$$M_A \cdot 0 + 2\,M_B\,(0 + l) + M_C\,l = -6\,E\,I\,\frac{\delta_C - \delta_B}{l}$$

or with $\delta_C - \delta_B = \varDelta\,\delta$

$$2\,M_B + M_C = -\,\frac{6\,E\,I \cdot \varDelta\,\delta}{l^2} \qquad (9)$$

or related to C:

$$M_B \cdot l + 2\,M_C\,(l + 0) + M_D \cdot 0 = +6\,E\,I\,\frac{\delta_C - \delta_B}{l}$$

or $\delta_C - \delta_B = \varDelta\,\delta$

$$M_B + 2\,M_C = +\,\frac{6\,E\,I \cdot \varDelta\,\delta}{l^2}\,. \qquad (10)$$

For the beam which is built-in at one end only, it follows, if

a) the support C is freely rotatable: $M_C = 0$ and

$$M_B = -\,\frac{3\,E\,I \cdot \varDelta\,\delta}{l^2}\,, \qquad (11)$$

b) the support B is freely rotatable: $M_B = 0$

$$M_C = -\frac{3EI \cdot \Delta\delta}{l_2}.\tag{12}$$

In Eq. (9) to (12), $\Delta_C = \delta_B - \delta_B$ should be taken as an algebraic value, so that the negative sign in these equations does *not* mean that M_B and M_C are in principle negative!

C. Relationships between loading (q_x), shearing forces (Q_x), moments (M_x), angles of rotation (α, β) and deflection curve (y)

In this section, we will be dealing with the purely mathematical and hence differential geometric relationships between the above magnitudes.

1. Loading and shearing force

According to Fig. 5, let us take any span of a continuous structure. The span width $\overline{AB} = l$. The continuous, differential and integrable load function shall be $q_x = f(x)$.

Let the shearing forces A_r and B_l, as well as the support moments M_A and M_B be assumed as determined by some means.

From the equilibrium condition $\Sigma V = 0$, it follows, for the shearing force at x, that:

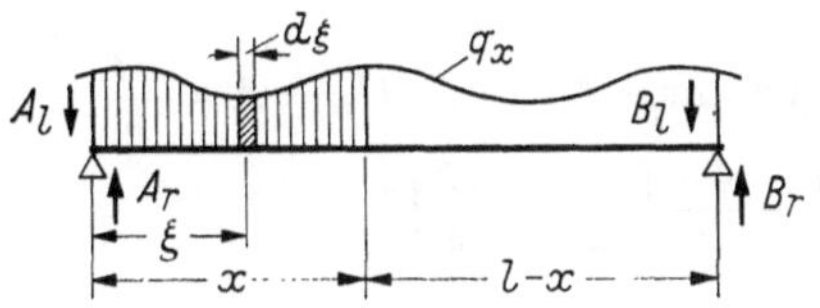

Fig. 5

$$Q_x = A_r - \int_0^x q_\xi \, d\xi,\tag{13}$$

in which ξ is the variable of integration and x any desired limit of integration which is fixed, however, during the process of integration.

Now, as q_r and q_ξ are the same functions in x as in ξ, Eq. (13) can also be more simply written, thus:

$$Q_x = A_r - \int q_x \, dx.\tag{14}$$

In differentiation of Eq. (14), the arbitrary constant A_r disappears and we obtain:

$$\boxed{\frac{dQ_x}{dx} = -q_x}.\tag{15}$$

Eq. (15) can be expressed in words by the following *sentence*:

> The derivation of the shearing force function is equal to the negative function of the load.

All the differential geometric relations between these two functions arise from this knowledge.

Let us quote two examples:

Example 1:

Simply-supported beam with a symmetric triangular load function. For reasons of symmetry, the support force A must equal $B = \dfrac{q\,l}{4}$. Generally, for $0 \leq x \leq -\dfrac{l}{2}$ the following applies:

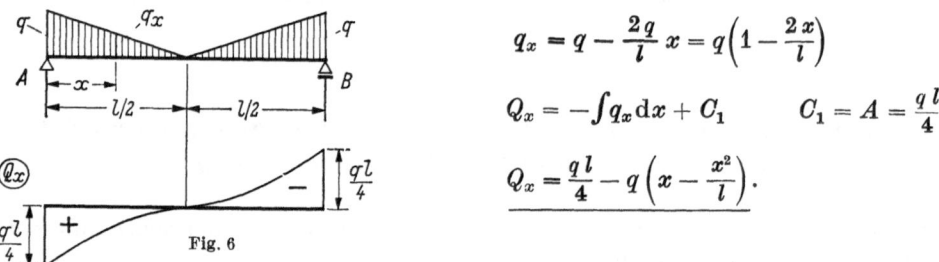

$$q_x = q - \frac{2\,q}{l}\,x = q\left(1 - \frac{2\,x}{l}\right)$$

$$Q_x = -\int q_x \mathrm{d}x + C_1 \qquad C_1 = A = \frac{q\,l}{4}$$

$$\underline{Q_x = \frac{q\,l}{4} - q\left(x - \frac{x^2}{l}\right).}$$

Fig. 6

The function of shearing forces is a quadratic parabola with a horizontal tangent at $x = \dfrac{l}{2}$, because the derivation is nil at that point. Maximum slope of the diagram is at the support, because it is there that the derivation has its maximum value ($q_x = q$). It is this which gives the form and character of the curve.

The validity of the function $Q_x = f\,(x)$ is limited to the range $0 \leq x \leq \dfrac{l}{2}$. The rest of the curve s provided by the conditions of symmetry.

Example 2:

Simply-supported beam under parabolic load

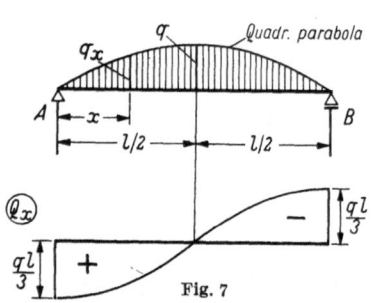

$$q_x = \frac{4\,q}{l^2}\,(l\,x - x^2)$$

$$Q_x = -\int q_x \mathrm{d}x + C_1 \qquad C_1 = A = \frac{q\,l}{3}$$

$$Q_x = \frac{q\,l}{3} - \frac{4\,q}{l^2}\left[\frac{l\,x^2}{2} - \frac{x^3}{3}\right].$$

Fig. 7

The shearing force function is a cubic parabola with horizontal tangents ($q_x = 0$) for $x = 0$ and $x = 1$ and with a point of inflection at $x = \dfrac{l}{2}$, because the differential curve q_x is at its maximum here.

2. Shearing forces and moments

According to Fig. 5, the moment at x is:

$$M_x = A_r\,x - \int\limits_0^x q_\xi\,(x - \xi)\,\mathrm{d}\xi + M_A.$$

Here, we see clearly that M_A assumes the value of an arbitrary constant.

Development of the equation for M_x gives us:

$$M_x = A_r\,x - x\int\limits_0^x q_\xi\,\mathrm{d}\xi + \int\limits_0^x q_\xi\,\xi\,\mathrm{d}\xi + M_A.$$

The differentiation of this equation to x gives us:

$$\frac{\mathrm{d}\,M_x}{\mathrm{d}\,x} = A_r - \int\limits_0^x q_\xi\,\mathrm{d}\xi.$$

According to Eq. (13), $A_r - \int_0^x q_\xi \, \mathrm{d}\xi = Q_x$, and it follows that:

$$\boxed{\frac{\mathrm{d}M_x}{\mathrm{d}x} = Q_x}. \tag{16}$$

Statement: | The derivation of the moment function is the shearing force function
or
the integral function of the shearing force function is the moment function.

If, then, we take $Q_x = -\int q_x \mathrm{d}x + C_1$, then M_x can be represented as

$$M_x = \int(-\int q_x \, \mathrm{d}x + C_1) \, \mathrm{d}x + C_2. \tag{17}$$

The moment function of a beam can therefore be ascertained by double integration of the negative load function, the arbitrary constants representing the conditions at the support of the span. It is therefore generally true that:

$$\boxed{\frac{\mathrm{d}^2 M}{\mathrm{d}x^2} = -q_x}. \tag{18}$$

Eq. (18) can be regarded as a differential equation of the moment diagram of a beam which is loaded by vertical loads. As we know from general differential and integral calculus, the funicular polygon drawn according to the loads represents the graphic integration of Eq. (18), the closing line representing fulfilment of the conditions of supports, in other words the ascertainment of the integration constants.

Since, in practice, the load functions are almost exclusively rational integral functions, in other words $q_x = a_0 + a_1 x + a_2 x^2 + \cdots + a_n x^n$, this function rises by one degree with every integration. Therefore, the shearing force function is higher than the load function by one degree and the moment function by two degrees.

Here again are two examples:

Example 1:

Simply-supported beam under uniformly distributed load.

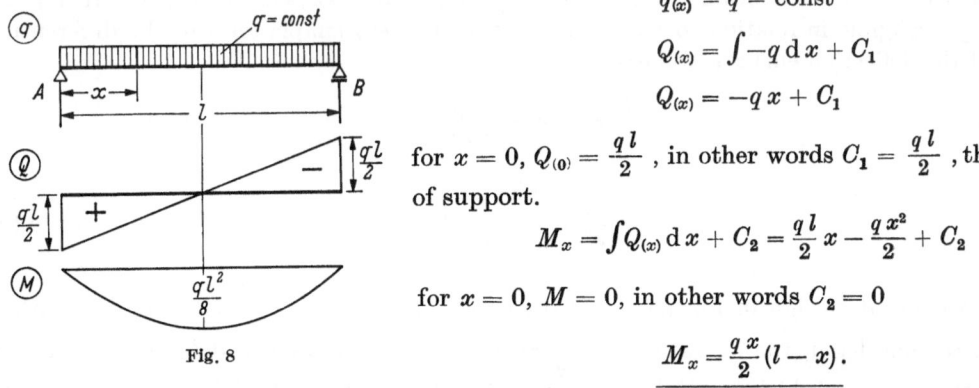

$$q_{(x)} = q = \text{const}$$
$$Q_{(x)} = \int -q \, \mathrm{d}x + C_1$$
$$Q_{(x)} = -qx + C_1$$

for $x = 0$, $Q_{(0)} = \frac{ql}{2}$, in other words $C_1 = \frac{ql}{2}$, the reaction of support.

$$M_x = \int Q_{(x)} \, \mathrm{d}x + C_2 = \frac{ql}{2}x - \frac{qx^2}{2} + C_2$$

for $x = 0$, $M = 0$, in other words $C_2 = 0$

$$\underline{M_x = \frac{qx}{2}(l - x)}.$$

Fig. 8

The function $Q_{(x)}$ has a constant slope, since its 1st derivation, the load, is constant, which means it is of the 1st degree, and passes through zero at the point of symmetry. Therefore, $M_{(x)}$-function is a 2nd degree parabola and, at $x = \frac{l}{2}$, has its maximum $M_{\max} = \frac{ql^2}{8}$.

Example 2:

Here, we take again the same example as was quoted for $Q_{(x)}$.

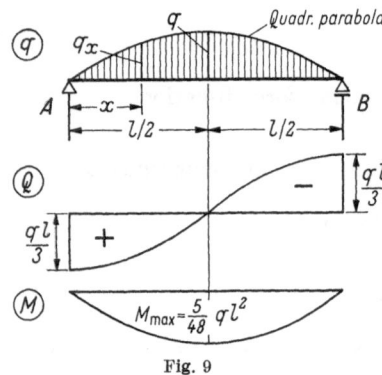

Fig. 9

$$q_{(x)} = \frac{4q}{l^2}(lx - x^2).$$

We saw, that

$$Q_{(x)} = \frac{ql}{3} - \frac{4q}{l^2}\left(\frac{lx^2}{2} - \frac{x^3}{3}\right).$$

If the moment at x is to be determined as a reduction of all forces on the left of the cross-section, then this automatically leads to integration, because both the hatched area and also the position of the centre of gravity cannot be ascertained otherwise. Therefore, we intend simply applying the rule that $M_{(x)} = \int Q_{(x)}\,dx + C_2$:

$$M_{(x)} = \frac{ql}{3}x - \frac{4q}{l^2}\left[\frac{lx^3}{6} - \frac{x^4}{12}\right] + C_2 = \frac{q}{3}\left[lx - \frac{1}{l^2}(2lx^3 - x^4)\right] + C_2.$$

Since all terms of this polynomial contain x, the arbitrary constant C_2 must be the moment for $x = 0$, in other words again a condition of support. $C_2 = M_{x=0} = 0$. The bending moment diagram is a quartic parabola with the maximum at $x_m = \frac{l}{2}$ of the magnitude $M_{\max} = \frac{5}{48}ql^2$.

3. The deflection curve and the angles of rotation at the ends

As the reader will know, the differential equation of the deflection curve of a beam is as follows:

$$\frac{1}{\varrho} = \frac{M}{EI}, \tag{19}$$

in which ϱ is the radius of curvature of the deflection curve at any one point. The following generally apply for ϱ as the radius of curvature of a function $f(x)$:

$$\varrho = \frac{(\sqrt{1 + f'(x)^2})^3}{f''(x)} \quad \text{oder} \quad \frac{1}{\varrho} = \frac{f''(x)}{(\sqrt{1 + f'(x)^2})^3}.$$

Since the deflection curves of beams encountered in practice are always extremely flat, then $f'(x)$ is absolutely negligible in relation to 1 and so we have the very simple form of the differential equation of the deflection curve as follows

$$\frac{1}{\varrho} \approx f''(x) = \frac{d^2 y}{dx^2}$$

as

$$\boxed{\frac{d^2 y}{dx^2} = -\frac{M}{EJ}}. \tag{20}$$

As regards the negative sign in Eq. (20), this is required more by convention than by mathematics. On the one hand, the curvature $k = \frac{1}{\varrho}$ of a curve pre-supposes a definite direction of revolution of the curve (which would lead to a positive sign) and, on the other hand, we want, for practical purposes, with positive moments, always to define positive ordinates of the deflection curve (deflections). This convention regarding sign also arises in a similar fashion in Eq. (18) where downwardly directed loads create likewise defined moments. In order to preserve this convention which is very important in practice, we must keep to the negative sign.

Differentiation of Eq. (20) gives us:

$$\frac{\mathrm{d}^3 y}{\mathrm{d} x^3} = -\frac{Q_x}{E I} \qquad (21)$$

and

$$\frac{\mathrm{d}^4 y}{\mathrm{d} x^4} = q_x \quad . \qquad (22)$$

According to Eq. (22), the function of the deflection curve can be obtained by quadrupled integration of the load function, the general rules of integration (regions of continuity) needing to be observed. We then have four integration constants C_1, C_2, C_3 and C_4, of which we are already familiar with C_1 and C_2 as shearing forces at supports and moments about points of support respectively. Therefore, it only remains to carry out a double integration of Eq. (20). We obtain

$$\frac{\mathrm{d} y}{\mathrm{d} x} = -\int \frac{M}{E I} \mathrm{d} x + C_3 \qquad (23)$$

and

$$y = \int \left(C_3 - \int \frac{M}{E I} \mathrm{d} x \right) \mathrm{d} x + C_4 . \qquad (24)$$

If we designate the *reduced moment diagram* by M/EI then, as a comparison with Eq. (13) will readily show, $\mathrm{d} y/\mathrm{d} x$ is the *imaginary function of shearing forces* for the *imaginary loading* given by the reduced moment diagram. The arbitrary constant C_3 is then the *reaction of support* arising from this imaginary loading. On the other hand, however, $\mathrm{d} y/\mathrm{d} x$ is the 1st derivation of the function of deflection curve and, as such, gives at any point the tangent slope of this curve. Since the angles arising are very small, φ may be taken as $\varphi \approx \tan \varphi$. In the case of rational integral functions of load, M is similarly one, so that for $I = \mathrm{const}$ in $\int \frac{M}{E I} \mathrm{d} x + C_3$, the value C_3 is the only term without x. C_3 therefore represents the angle formed by the deflection curve at the support with the axis of the beam, for $x = 0$, which we usually refer to as the angle of rotation at the end.

Since, in the tables, it is always assumed that I is constant, the values for the angles of rotation at the end always appear as multiplied by EI.

Since the integral of a rational integral function is always still rational, then the deflection curve $y = f(x)$ must in turn abide by a rational integral function at $E I = \mathrm{const}$ for every rational integral load function $q = f(x)$.

The following example should explain the value of the above relationships in practical application:

Example:

Simply-supported beam with triangular load

Section of continuity of the load: $0 \le x \le \frac{l}{2}$.

$x = \frac{l}{2}$ is the **axis of symmetry**.

1. Load function: $\qquad q_x = \frac{2 q}{l} x$

2. Shearing forces: $\qquad Q_x = -\int q_x \, \mathrm{d} x + C_1$

$\qquad C_1 = \text{reaction of support} \quad A = \frac{q l}{4}$

$\qquad Q_x = \frac{q l}{4} - \frac{q x^2}{l}$ (quadratic parabola).

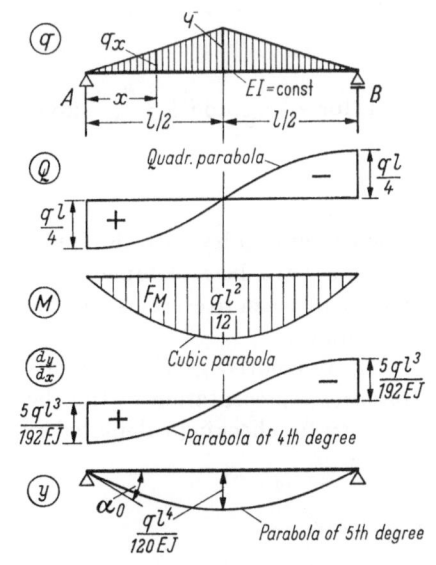

Fig. 10

3. Moments:
$$M_x = \int Q_x \, \mathrm{d}x + C_2 \qquad C_2 = M_{x=0} = 0$$

$$M_x = \frac{q\,l\,x}{4} - \frac{q\,x^3}{3\,l} \quad \text{(cubic parabola)}.$$

$$M_{\max} = \frac{q\,l^2}{12} \quad \text{at} \quad x = \frac{l}{2}.$$

Area of the bending moment diagram F_M:

$$F_M = 2 \int_0^{l/2} M_x \, \mathrm{d}x = 2\,q \int_0^{l/2} \left(\frac{l\,x}{4} - \frac{x^3}{3\,l} \right) \mathrm{d}x = 2\,q \left(\frac{l\,x^2}{8} - \frac{x^4}{12\,l} \right)_0^{l/2} = \frac{5\,q\,l^3}{96}.$$

4. Function:

$$\frac{\mathrm{d}y}{\mathrm{d}x} = \frac{1}{EI} \int -M_{(x)} \, \mathrm{d}x + C_3 = \frac{q}{24\,EI\,l} \left(2\,x^4 - 3\,l^2\,x^2 \right) + C_3.$$

Für $x = 0$, C_3 becomes the *reaction of the support* of the *imaginary loading* $\dfrac{M}{EI}$ i. e.

$$C_3 = \frac{F_M}{2\,EI} \qquad\qquad C_3 = \frac{5\,q\,l^3}{192\,EI}$$

and hence

$$\frac{\mathrm{d}y}{\mathrm{d}x} = \frac{q}{192\,EI\,l} \left(16\,x^4 - 24\,l^2\,x^2 + 5\,l^4 \right).$$

5. Deflection curve:

A further integration gives us the equation of the deflection curve, valid in the section $0 \le x \le \dfrac{l}{2}$, as:

$$y = \frac{q}{960\,EI\,l} \left(16\,x^5 - 40\,l^2\,x^3 + 25\,l^4\,x \right) + C_4.$$

Since, for $x = 0$, the deflection at the support is nil, then $C_4 = 0$.

If we now introduce the ratio $\xi = \dfrac{x}{l}$, then the equation for the deflection curve becomes

$$y = \frac{q\,l^4}{960\,EI} \left(16\,\xi^5 - 40\,\xi^3 + 25\,\xi \right),$$

which, for $x = \dfrac{l}{2}$ and $\xi = \dfrac{1}{2}$, gives the value of maximum deflection

$$y_{\max} = \frac{q\,l^4}{120\,EI}.$$

Since these tables are devised in the main for calculating continuous beams, formulae fo deflection have been completely disregarded. Values which could have been derived for th single-span beam are useless in many cases for load combinations and are too inaccurate eve for estimates.

The differential geometric relationships shown in this section allow, if the load is known, qualitative representation of the shearing forces and moments and so represent an excellent op portunity for checking calculated values. This is particularly true for frame constructions.

D. Examples of application of the tabular values for the singlespan beam under vertical loading

Example 1:

Given: a two-span beam, per Fig. 11a−c. EI = const.

The beam is to be calculated by the equation of three moments.

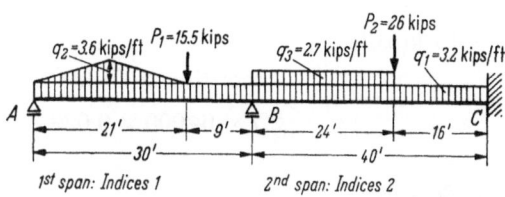

Fig. 11a

$$140\,M_B + 40\,M_C = -6\,(\beta_{01} + \alpha_{02})$$

$$40\,M_B + 80\,M_C = -6\,\beta_{02}.$$

If we term

$$a_B = -6\,(\beta_{01} + \alpha_{02})$$

and

$$a_C = -6\,\beta_{02}$$

then
$$M_B = \dfrac{\begin{vmatrix} a_B & 11 \\ a_C & 22 \end{vmatrix}}{\begin{vmatrix} 42 & 11 \\ 11 & 22 \end{vmatrix}} \quad \text{and} \quad M_C = \dfrac{\begin{vmatrix} 42 & a_B \\ 11 & a_C \end{vmatrix}}{\begin{vmatrix} 42 & 11 \\ 11 & 22 \end{vmatrix}}.$$

The values a_B and a_C are obtained as follows:

β_{01}: This is the angle of rotation at the supports, multiplied by EI, for the right support (hence β) on a statically determined single-span beam (hence Index $_0$) in the 1st span (hence the Index $_1$).

1. From the load $q_1 = 3.2$ kips/ft:

   According to Table 1, p. 29, $n = 1$

   $3.2 \times 27\,000 \times 0.042 \quad = \qquad\qquad 3628.8$

2. From the load $q_2 = 3.6$ kips/ft:

   According to Table 10, p. 38, $n = 0.7$

   $3.6 \times 27\,000 \times 0.017 \quad = \qquad\qquad 1652.4$

3. From the load $P_1 = 15.5$ kips:

   According to Table 27, p. 83

   $15.5 \times 900 \times 0.060 \quad = \qquad\qquad \underline{837.0}$

   $\beta_{01} = \qquad \underline{\underline{6118.2}}$

2*

α_{02}: This is the angle of rotation at the supports, multiplied by EI, for the left-hand **support** (hence α) on a statically determined single-span beam (hence the Index $_0$) in the 2nd **span** (hence the Index $_2$).

1. From the load $q_1 = 3.2$ kips/ft:

According to Table 1, p. 29, $n = 1$

$3.2 \times 64000 \times 0.042 \quad = \qquad\qquad 8601.6$

2. From the load $q_3 = 2.7$ kips/ft:

According to Table 1, p. 29, $n = 0.6$

$2.7 \times 64000 \times 0.029 \quad = \qquad\qquad 5011.2$

3. From the load $P_2 = 26$ kips:

According to Table 27, p. 83, $n = 0.6$

$26 \times 1600 \times 0.056 \qquad = \qquad\qquad \underline{2329.6}$

$$\alpha_{02} = 15942.4$$

β_{02}: This is the angle of rotation at the supports, multiplied by EI, for the right-hand support of the statically determined single-span beam, in the 2nd span.

1. From the load $q_1 = 3.2$ kips/ft:

According to Table 1, p. 29, $n = 1$

$3.2 \times 64000 \times 0.042 \quad = \qquad\qquad 8601.6$

2. From the load $q_2 = 2.7$ kips/ft:

According to Table 1, p. 29, $n = 0.6$

$2.7 \times 64000 \times 0.025 \quad = \qquad\qquad 4320.0$

3. From the load $P_2 = 26$ kips:

According to Table 27, p. 83, $n = 0.6$

$26 \times 1600 \times 0.064 \qquad = \qquad\qquad \underline{2662.4}$

$$\beta_{02} = 15584.0$$

From these values, it follows that:

$$a_B = -6\,(6118.2 + 15942.4) = -132363.2$$

$$a_C = -6 \times 15584.0 \qquad\qquad = -\;93504.0$$

and further:

$$M_B = \frac{\begin{vmatrix} -132363.2 & 40 \\ -93504.0 & 80 \end{vmatrix}}{\begin{vmatrix} 140 & 40 \\ 40 & 80 \end{vmatrix}} = -\frac{6848896}{9600} = \underline{-712.6 \text{ ft-kips}}$$

$$M_C = \frac{\begin{vmatrix} 140 & -132363.2 \\ 40 & -93504.0 \end{vmatrix}}{\begin{vmatrix} 140 & 40 \\ 40 & 80 \end{vmatrix}} = -\frac{7796032}{9600} = \underline{-812.1 \text{ ft-kips}}$$

It is intended now to show also the diagrams for bending moments and shearing forces:

The M-Diagrams:

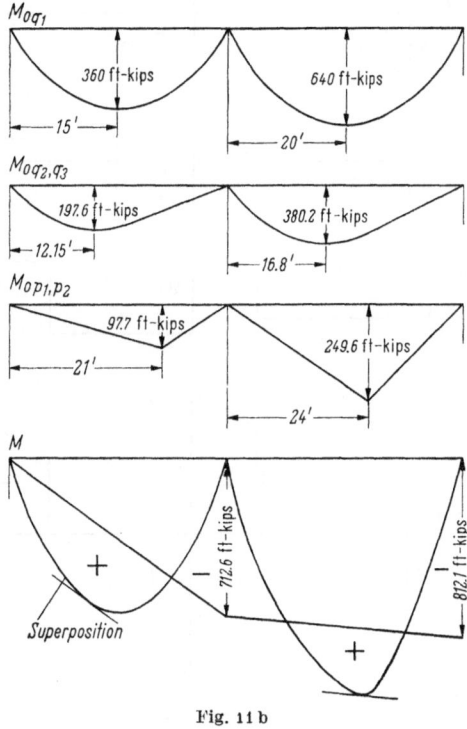

Fig. 11 b

1st Span:

a) Due to q_1:

$M_{0\,max} = k_7 q l^2$

Per Table 1, p. 29, $\quad n = 1$

$M_{0\,max} = 0.125 \times 3.2 \times 900 \quad = 360$ ft-kips

at $\quad x_m = k_8 \; l = 0.5 \times 30' \quad = 15$ ft

b) Due to q_2:

$M_{0\,max} = k_7 q l^2$

Per Table 10, p. 38, $\quad n = 0.7$

$M_{0\,max} = 0.061 \times 3.6 \times 900 \quad = 197.6$ ft-kips

at $\quad x_m = 0.407 \times 30' \quad = 12.21$ ft

c) Due to $P_1 = 15.5$ kips: $\quad$ Per Table 27, p. 83 $\quad n = 0.7$

$M_{0\,max} = 0.21 \times 15.5 \times 30 \quad = 97.7$ ft-kips

at $\quad x_m = nl = 0.7 \times 30 \quad = 21$ ft

2nd Span:

a) Due to q_1:

Per Table 1, $\quad\quad n = 1$

$M_{0\,max} = 0.125 \times 3.2 \times 1600 \quad = 640$ ft-kips

at $\quad x_m = 0.5 \times 40' \quad = 20$ ft

b) Due to q_3:

Per Table 1, $\quad\quad n = 0.6$

$M_{0\,max} = 0.088 \times 2.7 \times 1600 \quad = 380.2$ ft-kips

at $\quad x_m = 0.42 \times 40' \quad = 16.8$ ft

c) Due to $P_2 = 26$ kips: $\quad$ Per Table 27

$M_{0\,max} = 0.24 \times 26 \times 40 \quad = 249.6$ ft-kips

The Q-Diagrams:

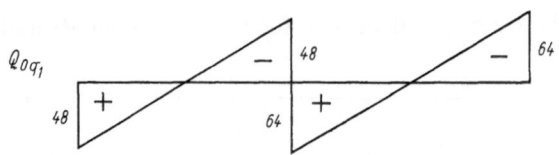

Q_{0q_1}

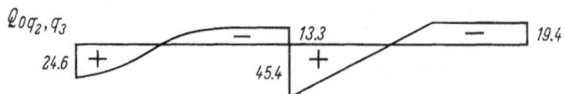

Q_{0q_2}, q_3

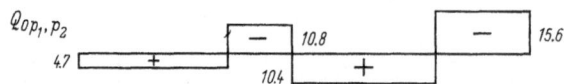

Q_{0p_1}, p_2

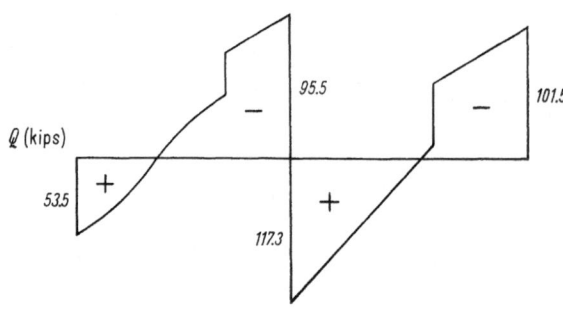

Q (kips)

Fig. 11 c

1st Span:

    a) Due to q_1: $Q_{01} = k_3 q_1 l_1$; $Q_{02} = k_4 q_1 l_1$.

                        Per Table 1, $n = 1$

                        $Q_{01} = 0.5 \times 3.2 \times 30$ $= 48.0$ kips

                        $Q_{02} = Q_{01}$

    b) Due to q_2:   Per Table 10, $n = 0.7$

                        $Q_{01} = 0.228 \times 3.6 \times 30$ $= 24.6$ kips

                        $Q_{02} = 0.123 \times 3.6 \times 30$ $= 13.3$ kips

    c) Due to P_1:   Per Table 27, $n = 0.7$

                        $Q_{01} = 0.3 \times 15.5$ $=\ 4.7$ kips

                        $Q_{02} = 0.7 \times 15.5$ $= 10.8$ kips

2nd Span:

    a) Due to q_1:   Per Table 1

                        $Q_{01} = Q_{02} = 0.5 \times 3.2 \times 40$ $= 64.0$ kips

    b) Due to q_3:   Per Table 1 $n = 0,6$

                        $Q_{01} = 0.42 \times 2.7 \times 40$ $= 45.4$ kips

                        $Q_{02} = 0.18 \times 2.7 \times 40$ $= 19.4$ kips

c) Due to P_2:   Per Table 27 $n = 0.6$

$$Q_{01} = 0.4 \times 26 \qquad\qquad = 10.4 \text{ kips}$$

$$Q_{02} = 0.6 \times 26 \qquad\qquad = 15.6 \text{ kips}$$

For qualitative determination of the Q_0-diagrams, the remarks given in section B apply. Since $Q_x = -\int q_x \, dx + C$, it can immediately be established that the Q-diagram,

due to q_1: is a linear function with a zero at every $l/2$,

due to q_2: gives two parts of quadratic parabolae each being limited by an horizontal tangent ($q_{x2} = 0$) with its maximum slope at the symmetry (inflection) point at $x = 10.5'$ (when q_{x2} is at its maximum) and remains constant in the section $21' \leq x \leq 30'$ ($= Q_{02}$), because in that case $q_{x2} = 0$. The zero point is at $x_m = 12.15$ ft according to $\dfrac{dM_x}{dx} = Q_x$. Where $Q_x = 0$, $M_x \to M_{\max}$.

Due to q_3: is a linear function with the zero at $x_m = 21$ ft, again according to $\dfrac{dM_x}{dx} = Q_x$.

In the final analysis, for superposition, the influence of the moments must be taken into account, which gives us an additive constant, viz

for the first span $Q_{1M} = -\dfrac{712.6}{30} = -23.8 \text{ kips}$

for the second span $Q_{2M} = -\dfrac{812.1 - 712.6}{40} = -2.5 \text{ kips.}$

The shearing forces at the supports are therefore:

Right of A: $48.0 + 24.6 + 4.7 - 23.8 = 53.5 \text{ kips},$

left of B: $-48.0 - 13.3 - 10.8 - 23.8 = -95.5 \text{ kips},$

right of B: $64.0 + 45.4 + 10.4 - 2.5 = 117.3 \text{ kips},$

left of C: $-64.0 - 19.4 - 15.6 - 2.5 = -101.5 \text{ kips.}$

The reactions of supports:

$$A = 53.5 \text{ kips}$$

$$B = 95.5 + 117.3 = 212.8 \text{ kips}$$

$$C = 101.5 \text{ kips.}$$

Final check:

The total of the support reactions

$$53.5 + 212.8 + 101.5 = 367.8 \text{ kips}$$

must be equal to the total of all loads:

$$70 \times 3.2 + 21 \times 0.5 \times 3.6 + 24 \times 2.7 + 15.5 + 26.0 = 368.1 \text{ kips.}$$

The mistake caused by rounding-off figures, amounts to

$$\Delta = \frac{0.3}{368,1} = 0.00081 = 0.81^0/_{00}.$$

Example 2:

Problem: to ascertain the moments about points of support for the following triple-span beam, using the Cross method.

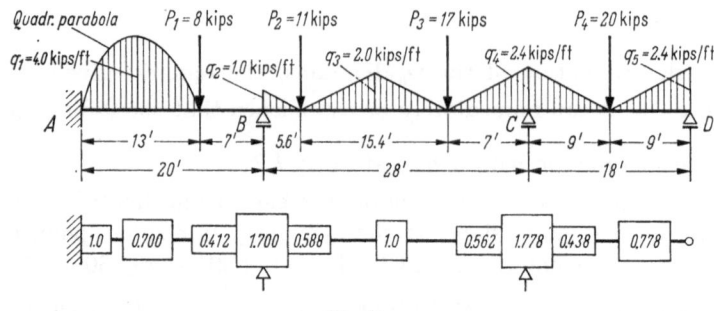

Fig. 12

The stiffnesses s:

Let the moments of inertia be

$$I_2 : I_1 = 2$$
$$I_2 : I_3 = 2.5 .$$

Related to the middle span, with $s_2 = 1$, we then have

$$s_1 = \frac{28}{20} \times \frac{1}{2} = 0.7; \qquad s_2 = 1; \qquad s_3 = \frac{28}{18} \times \frac{1}{2.5} \times 0.75 = 0.778 .$$

Ascertaining the fixed-end moments:

1st Span:

	M_1 (ft-kips)	$-M_2$ (ft-kips)
a) From q_1 per Table 16, with $n = \frac{13}{20} = 0.65$:		
$M_1 = k_1 q l^2 = 0.055 \times 4.0 \times 400 \quad =$	88.0	
$M_2 = k_2 q l^2 = 0.031 \times 4.0 \times 400 \quad =$		49.6
b) From P_1 per Table 27, with $n = 0.65$:		
$M_1 = k_1 P l = 0.080 \times 8 \times 20 \quad =$	12.8	
$M_2 = k_2 P l = 0.148 \times 8 \times 20 \quad =$		23.7
Total, 1st Span:	100.8	83.3

2nd Span:

a) From q_2, per Table 5, with $n = \frac{5.6}{28} = 0.20$:		
$M_1 = k_1 q l^2 = 0.005 \times 1.0 \times 784$	3.9	
$M_2 = k_2 q l^2 = 0.001 \times 1.0 \times 784 \quad =$		0.8
b) From q_3, per Table 24, $m = 0.20$, $n = \frac{15,4}{28} = 0.55$		
$M_1 = 0.034 \times 2.0 \times 784 \quad =$	53.3	
$M_2 = 0.031 \times 2.0 \times 784 \quad =$		48.6
Carry forward:	57.2	49.4

	M_1 (ft-kips)	$-M_2$ (ft-kips)
Brought forward, 2nd Span:	57.2	49.4

c) From P_2, per Table 27, $n = 0.20$

$\qquad k_1 = 0.128;\ k_2 = 0.032$

$\qquad M_1 = 0.128 \times 11 \times 28 \qquad\qquad = \qquad 39.4$

$\qquad M_2 = 0.032 \times 11 \times 28 \qquad\qquad = \qquad\qquad\qquad 9.9$

From P_3, per Table 27, $n = \dfrac{21}{28} = 0.75$

$\qquad k_1 = 0.047;\ k_2 = 0.141$

$\qquad M_1 = 0.047 \times 17 \times 28 \qquad\qquad = \qquad 22.4$

$\qquad M_2 = 0.141 \times 17 \times 28 \qquad\qquad = \qquad\qquad\qquad 67.1$

d) From q_4, per Table 5, $n = \dfrac{7}{28} = 0.25$

$\qquad k_1 = 0.001;\ k_2 = 0.008$

$\qquad M_1 = 0.001 \times 2.4 \times 784 \qquad\qquad = \qquad 1.9$

$\qquad M_2 = 0.008 \times 2.4 \times 784 \qquad\qquad = \qquad\qquad\qquad 15.1$

	M_1	$-M_2$
Total, 2nd Span:	120.9	141.5

3rd Span:

a) From q_4 and q_5. Symmetrie!

$\qquad$ Per Table 7, p. 35, $n = 0.5$

$\qquad k_1^0 = 0.047$

$\qquad M_1^0 = 0.047 \times 2.4 \times 324 \qquad\qquad = \qquad 36.5$

b) From P_4, per Table 27; $n = 0.5$

$\qquad M_1^0 = 0.188 \times 20 \times 18 \qquad\qquad = \qquad 67.7$

$\qquad\qquad\qquad$ Total, 3rd Span: $\quad = \qquad 104.2$

-104.9	-127.2
$-\ \ 0.5$	$+\ \ \ 2.2$
$-\ \ 5.6$	$-\ \ \ 4.0$
$-\ 15.5$	$+\ 27.1$
$-\ 83.3$	$-\ 11.0$
	-141.5

The moments about points of support, as sought, are:

$M_A = -\ 90\ $ ft-kips

$M_B = -\ 104.9$ ft-kips

$M_C = -\ 127.2$ ft-kips

$M_D = 0$.

$+100.8$	$+120.9$	$+104.2$
$-\ \ 7.8$	$-\ 22.1$	$+\ 21.2$
$-\ \ 2.8$	$+\ 13.5$	$+\ \ 1.8$
$-\ \ 0.2$	$-\ \ 7.9$	$+127.2$
$+\ 90.0$	$+\ \ 1.1$	
	$-\ \ 0.6$	
	$+104.9$	

As this example clearly shows, where are several distributed loads in the same span, the square of the spans always recurs.

If, therefore, a moment M_1 is composed of:

$$M_1 = k'\, q'\, l^2 + k_1''\, q''\, l^2 + \cdots + k_1^{(n)}\, q^{(n)}\, l^2 ,$$

then we can determine more simply

$$M_1 = l^2\, (k_1'\, q' + k_1''\, q'' + \cdots + k_1^{(n)}\, q^{(n)}) .$$

For concentrated loads, the procedure is similar, in that we determine

$$M_1 = k_1'\, P'\, l + k_1''\, P''\, l + \cdots + k_1^{(n)}\, P^{(n)}\, l$$

$$M_1 = l\, (k_1'\, P' + k_1''\, P'' + \cdots + k_1^{(n)}\, P^{(n)}) .$$

For this, please study the next example:

Example 3:

Determining the FEM's of a triple-span beam with one fixed end.

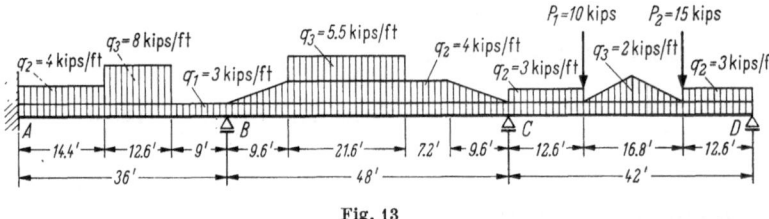

Fig. 13

Calculation is easiest from a Table:

Span	Load	Tab.No.	m	n	k_1 or k_1^0	$k_{1i}\, q_i$	k_2	$k_{2i}\, q_i$	l	l^2
1	$q_1 = 3.0$	1	0	1.0	0.083	0.249	0.083	0.249	36.0	1296
	$q_2 = 4.0$	1	0	0.4	0.044	0.176	0.015	0.060		
	$q_3 = 8.0$	22	0.4	0.35	0.035	0.280	0.047	0.376		
					$\sum k_{1i}\, q_i = 0.705$		$\sum k_{2i}\, q_i = 0.685$			

$M_1 = 1296 \times 0.705 = 913.7$ ft-kips. $\qquad M_2 = 1296 \times 0.685 = 887.8$ ft-kips.

Span	Load	Tab.No.	m	n	k_1 or k_1^0	$k_{1i}\, q_i$	k_2	$k_{2i}\, q_i$	l	l^2
2	$q_1 = 3.0$	1	0	1.0	0.083	0.249	0.083	0.249	48	2304
	$q_2 = 4.0$	13	0.2		0.077	0.308	0.077	0.308		
	$q_3 = 5.5$	22 A	0.2	0.45	0.058	0.319	0.043	0.237		
					$\sum k_{1i}\, q_i = 0.876$		$\sum k_{2i}\, q_i = 0.794$			

$M_1 = 2304 \times 0.876 = 2018.3$ ft-kips. $\qquad M_2 = 2304 \times 0.794 = 1829.4$ ft-kips.

Span	Load	Tab.No.	m	n	k_1 or k_1^0	$k_{1i}\, q_i$	k_2	$k_{2i}\, q_i$	l	l^2
3	$q_1 = 3.0$	1	0	1.0	0.125	0.375			42	1764
	$q_2 = 3.0$	3		0.3	0.054	0.162				
	$q_3 = 2.0$	11		0.4	0.037	0.074				
					$\sum k_{1i}^0\, q_i = 0.611$					
	$P_1 = 10$	27		0.3	0.179	1.790				
	$P_2 = 15$	27		0.7	0.137	1.905				
					$\sum k_{1i}^0\, P_i = 3.695$					

$M_1^0 = 1764 \times 0.611 + 42 \times 3.695 = 1233$ ft-kips.

We should couple with this example a few remarks about accuracy:

If 5 decimal places are observed for the k values, the FEM's will be

in the 1st span: $M_1 = $ 917.5 ft-kips $M_2 = $ 900.2 ft-kips,

in the 2nd span: $M_1 = 2020.5$ ft-kips $M_2 = 1830.5$ ft-kips,

in the 3rd span: $M_1 = 1237.3$ ft-kips.

Comparing these values to those of the above table we find an average error

$$\underline{\Delta_{\text{aver.}} \sim 0.4\%}$$

which is sufficient for practical requirements.

E. The tables for determining the secondary moments due to prestressing
(Tables V-1 to V-5)

1. General remarks

In determining the prestressing of a beam, two items of data must, in principle, be established:

the prestressing force   and
the position of the prestressing tendon.

The prestressing force V is normally obtained from the dimensioning of the main cross-sections, where, for a given V, an eccentricity e of the prestressing tendon results. According to the position of the prestressing heads, we then have a connecting line between these main eccentricities (e. g. e_{m_1}, e_2 and e_{m_2} in Fig. 14) through the prestressing tendon itself, the character of the prestressing tendon being as far as possible in accordance with the diagram of maximum and minimum moments.

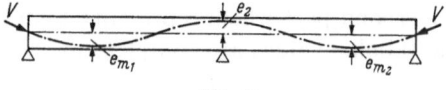

Fig. 14

The curves which meet this requirement are parabolae of a quadratic and higher order. Since, in practice, whith statically indeterminate systems, additional forces and moments (secondary moments) arise from the prestressing itself, modifying the original diagrams of maximum and minimum values caused by $g + p = q$, it is desirable to be able to have these moments available, as far as possible without a lot of arithmetical work. This requirement is met by Tables V-1 to V-5.

Where prestressing tendons have a curved character, friction losses during prestressing are inevitable. According to the nature of the prestressing tendon, the friction forces comply with another law or, if the basic friction law is the same, the coefficients of these laws will vary. Moreover, the specifications for considering friction vary from one case to another. By reason of these manifold considerations, the effect of friction has been omitted from these Tables.

The friction losses which can arise in a single-span, such as we are concerned with here, influence statically indeterminate moments $\bar{M}$ by roughly 3%.

The moment resulting from the prestressing force V which acts in a given cross-section with the eccentricity e_x, is $M_v = V e_x \cos \varphi$, in which φ is the slope of the tendon in cross-section x.

Normally, the prestressing tendons are very flat so that φ is very small and $\cos \varphi$ becomes approximately 1. In the formulae and Tables V-1 to V-5, $\cos \varphi$ is taken as $= 1$. The error acting on the statically indeterminate moments is generally less than 1%.

The errors due to omitting the friction and by assuming $\cos\varphi$ to be $= 1$ are small. Since the Tables are intended to give values for estimating the influence of prestressing on the statically indeterminate magnitudes, the simplification mentioned is entirely tolerable. On more important constructions, the effects of friction must of course be shown in the final calculation. The same applies to the relaxation of the prestressing tendons and also for the shrinkage and creep of the concrete.

2. Explanation of the tables

Table V-1 deals with the prestressing tendons which are laid according to a cubic parabola. All the necessary coefficients for a section of the prestressing tendon between its end-anchorage and the point of maximum eccentricity in the span (point of the prestressing cable with a horizontal tangent) are listed here.

If, then, we are to investigate the action of a prestressing tendon running over the entire span, it becomes necessary to add together two appropriate values, taking care only to see that the two partial lines appear in reverseimage in the tables (see the examples in Section 3!). For combinations with cubic parabola sections, the values given in Table V-4 may be used.

Table V-2, like Table V-1, deals with the cases of so-called Chapeau-cable which can additionally be provided over the intermediate supports, running according to quadratic parabolae, similarly in partial sections. Since these cables always relate to two bays of mostly different span, it is obvious that they should be treated in the respective sections; all the more since the distribution of moments or computation must be according to CLAPEYRON's theorem (basic values on different sides of the intermediate support!).

With Chapeau-cables, the error due to assuming $\cos\varphi = 1$ is substantially greater than under normal circumstances but, due to the fact that Capeau-cables are short, it does not assume vital importance as regards the overall span and its magnitudes.

Table V-3 relates to prestressing tendons to cubic parabolae as arise with intermediate supports, again in partial sections between two points with horizontal tangents. Every possibility is catered for by combining two partial sections (see also Table V-4 and V-5).

Table V-4 summarises the values for the prestressing tendons which comprise one part following a quadratic parabola and a second part which represents a cubic parabola. These cases occur particularly in end spans. The tabular values are complete in themselves and require no superposition.

Table V-5, finally, gives final values for inner spans, where two parts each follow a cubic parabola. These tables are grouped according to the ratios $r = e_1 : e_2$, in which e_1 signifies the eccentricity over the left support and e_2 the eccentricity over the right support.

3. Specimen applications

a) Example 1:

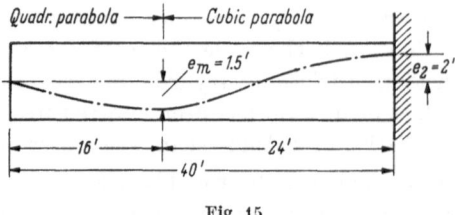

Fig. 15

Problem: to determine the moment $\overline{M}_2^0$ of a cable as shown in Fig. 15.

$$s = 16' = nl \qquad n = 0.4 \qquad m = e_m : (- e_2) = 0.75$$

According to Table V-1, for $n = 0.4$, $k_{12}^0 = 0.200$ and, accordingly the moment resulting from the left part of the prestressing tendon is

$$\bar{M}_{2l}^0 = 0.200\, e_m\, V\,.$$

According to Table V-3, now, for $n = 0.6$ and $m = 0.75$, the corresponding value k_{12}^0 must be determined. This follows as $k_{12}^0 = -0.347$, so that the moment corresponding to the right-hand part of the prestressing tendon is $\bar{M}_{2r}^0 = -0.347\, V\, e_2$.

The total moment $\bar{M}_2^0$ is then given as a superposition of the parts arising from the two pa sections:

$$\bar{M}_2^0 = 0.200\, e_m\, V - 0.347\, e_2\, V\,.$$

With $e_2 = \dfrac{1}{m}\, e_m$, $e_2 = 1.333\, e_m$ and

$$\bar{M}_2^0 = (0.200 - 1.333 \times 0.347)\, e_m\, V = -0.2625 e_m\, V\,.$$

This value obtained for the sections of the prestressing tendon, by applying the Tables could also have been obtained directly by applying Table V-4 to this case of combination:

Abiding by the designation used there $(s = nl)$, we find under $n = 0.60$ and $m = 0.75$ a value -0.262 for k_{12}^0, in other words the same coefficient as was obtained by dealing with the sections according to Tables V-1 and V-2; the error of 0.2 % results from rounding-off.

With $e_m = 1.5'$, the fixed end moment $\bar{M}_2^0$ becomes

$$\bar{M}_2^0 = -0.262 \times 1.5.\, V = -0.393\, V \quad \text{in ft-kips}\,.$$

b) Example 2:

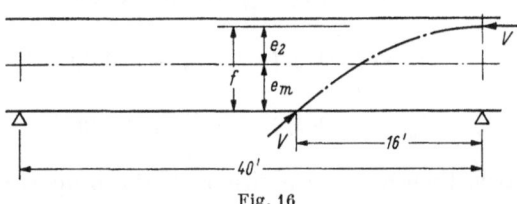

Fig. 16

A Chapeau-cable has a rise span $f = e_2 + e_m = 2'$ and, as shown in Fig. 16, is anchored at a distance of 16' from the support of a 40' long span.

To find: the angle of rotation at the support: $\bar{\beta}_0$ (value multiplied by EI).

By applying Table V-2, for $s = nl = 16'$, $n = 0.40$, we have

$$\bar{\beta}_0 = k_{15}\, V\, f\, l \qquad k_{15} = 0.200$$
$$\bar{\beta}_0 = 0.200 \times 2 \times 40 = 16\, V \quad \text{in ft}^2\text{-kips}\,.$$

c) Example 3:

For an inner span 73' long, the two angles of rotation $\bar{\alpha}_0$ and $\bar{\beta}_0$ must be determined and also the points at which the prestressing tendon intersects with the middle axis of the beam (points

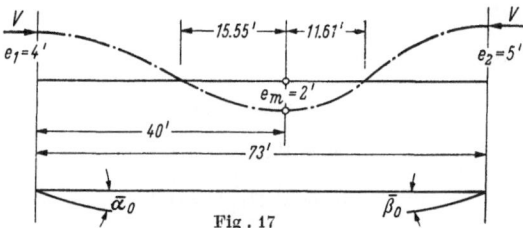

Fig. 17

with centric prestressing). The prestressing tendon should, as it is an inner span with which we are concerned, be composed of two parts according to cubic parabolae (see Fig. 17).

We will first solve the problem according to Table V-3, considering the prestressing tendon in two parts:

Left-hand part:

$$n_l = 40 : 73 = 0.55$$

$$m_l = e_m : (-e_1) = 0.5 \, .$$

For the left-hand part, we must apply Table V-3 mirror-symmetrically:

Proportion for $\bar{\alpha}_0$:   Due to the mirror symmetry, not k_{14} but k_{15}:

$$\bar{\alpha}_{0\,l} = k_{15} \, V \, e_1 \, l \qquad k_{15} = 0.145 \, .$$

Proportion for $\bar{\beta}_0$:   Due to the mirror symmetry, not k_{15} but k_{14}:

$$\bar{\beta}_{0\,l} = k_{14} \, V \, e_1 \, l \qquad k_{14} = -0.008 \, .$$

Right-hand part:

For this, the Table, according to the sketch of assumption, is applied directly and not mirror-symmetrically.

In this case, then:

$$n_r = 1 - 0.55 = 0.45$$

$$m_r = e_m : (-e_2) = 2 : 5 = 0.40 \, .$$

Proportion for $\bar{\alpha}_0$:

$$\bar{\alpha}_{0\,r} = k_{14} \, V \, e_2 \, l \qquad k_{14} = -0.002 \, .$$

Proportion for $\bar{\beta}_0$:

$$\bar{\beta}_{0\,r} = k_{15} \, V \, e_2 \, l \qquad k_{15} = 0.133 \, .$$

The total values are given then as superpositions of the proportions for the right and left-hand parts of the prestressing tendon:

$$\bar{\alpha}_0 = 0.145 \, V e_1 + 0.002 \, V \, e_2 \, l \qquad \text{and with } r = e_1 : e_2 = 0.80,$$

$$\bar{\alpha}_0 = (0.8 \times 0.145 + 0.002) \, V \, e_2 \, l \qquad \text{then } e_1 = 0.80 \, e_2, \text{ so that}$$

$$\underline{\bar{\alpha}_0 = 0.188 \, V \, e_2 \, l}$$

$$\bar{\beta}_0 = \quad 0.008 \, V \, e_1 + 0.133 \, V \, e_2 \, l \qquad r = e_1 : e_2 = 0.8$$

$$\bar{\beta}_0 = (0.133 - 0.8 \times 0.008) \, V \, e_2 \, l \qquad e_1 = 0.8 \, e_2$$

$$\underline{\beta_0 = 0.1266 \, e_2 \, l \, .}$$

The second method of solution, the direct method, is via Table V-5, where we can read off the final coefficients for this case:

Firstly, the indispensable parameters:

$$r = e_1 : e_2 = 0.8$$

$$n = 0.55$$

$$m = e_m : (e_2) = 0.4 \, .$$

For the ratio $r = 0.8$ we find in Table V-5h:

$$k_{14} = 0.118 \qquad k_{15} = 0.127 \, .$$

For the values sought, it follows that:

$$\bar{\alpha}_0 = 0.118 \ V \ e_2 \ l \quad \text{and} \quad \bar{\beta}_0 = 0.127 \ V \ e_2 \ l,$$

results which agree with those given in Table V-3, to within 0,3%.

The small inevitable difference is caused by rounding-off the values for the coefficients k.

It goes without saying that, as far as possible, one should always use the combination tables V-4 and V-5. In quite extreme cases which go beyond the scope of Tables V-4 and V-5, V-1 and V-3 will prove very satisfactory.

The second part of the problem: determining the points with eccentricity $e_x = 0$ (Position of the points of intersection of the prestressing tendon with the beam axis).

Left-hand part of the prestressing tendon: (To Table V-3b)

$$n_l = 0.55 \qquad m_l = 0.5 \qquad \bar{x}_{0l} = k_{16} \, l$$
$$k_{16} = 0.213$$
$$\bar{x}_{0l} = 0.213 \times 73 = 15.55'$$

measured from e_m towards the left-hand support.

Right-hand part of the prestressing tendon: (To Table V-3b)

$$n_r = 0.45 \qquad m_r = 0.4 \qquad \bar{x}_{0r} = k_{16} \, l$$
$$k_{16} = 0.159$$
$$\bar{x}_{0r} = 0.159 \times 73 = 11.61'$$

measured from e_m towards the right-hand support.

Table of the ratios of moments of inertia of T-cross-section to those of rectangular cross-section

$$J_T = \mu\,\frac{b\,H^3}{12} = \mu\,J_\square$$

$$\mu = \frac{J_T}{J_\square}$$

Table ratios μ:

H : d	B : b											
	1.5	2.0	2.5	3.0	3.5	4.0	4.5	5,0	5.5	6.0	6.5	7.0
3.0	1.210	1.370	1.500	1.605	1.700	1.780	1.845	1.910	1.965	2.018	2.065	2.110
3.5	1.205	1.365	1.495	1.605	1.695	1.775	1.845	1.910	1.970	2.020	2.065	2.110
4.0	1.195	1.355	1.485	1.595	1.685	1.770	1.840	1.905	1.965	2.015	2.060	2.105
4.5	1.185	1.340	1.470	1.580	1.675	1.760	1.835	1.900	1.960	2.010	2.055	2.100
5.0	1.175	1.330	1.455	1.565	1.655	1.745	1.820	1.885	1.945	2.000	2.045	2.095
5.5	1.170	1.315	1.440	1.545	1.640	1.725	1.805	1.870	1.935	1.985	2.035	2.085
6.0	1.165	1.305	1.425	1.530	1.625	1.710	1.785	1.855	1.915	1.970	2.030	2.070
6.5	1.155	1.290	1.410	1.515	1.605	1.690	1.765	1.835	1.895	1.950	2.005	2.055
7.0	1.145	1.275	1.385	1.495	1.590	1.670	1.745	1.815	1.875	1.935	1.985	2.035
7.5	1.140	1.265	1.380	1.480	1.570	1.650	1.725	1.795	1.855	1.915	1.965	2.015
8.0	1.135	1.255	1.365	1.465	1.555	1.635	1.705	1.775	1.835	1.895	1.945	1.995
8.5	1.130	1.245	1.350	1.450	1.535	1.615	1.685	1.755	1.815	1.875	1.925	1.975
9.0	1.125	1.235	1.340	1.435	1.515	1.595	1.670	1.735	1.800	1.850	1.905	1.955
9.5	1.120	1.230	1.330	1.420	1.505	1.580	1.650	1.715	1.780	1.835	1.885	1.935
10.0	1.115	1.220	1.315	1.410	1.490	1.565	1.635	1.700	1.760	1.815	1.865	1.915
10.5	1.110	1.215	1.310	1.395	1.475	1.550	1.620	1.680	1.740	1.795	1.850	1.900
11.0	1.105	1.205	1.300	1.385	1.460	1.535	1.600	1.665	1.725	1.775	1.830	1.880
11.5	1.100	1.200	1.290	1.370	1.450	1.520	1.585	1.650	1.705	1.760	1.815	1.865
12.0	1.100	1.195	1.280	1.360	1.435	1.505	1.570	1.635	1.690	1.745	1.795	1.845
12.5	1.095	1.190	1.265	1.350	1.425	1.495	1.550	1.620	1.675	1.730	1.780	1.825
13.0	1.090	1.185	1.260	1.340	1.415	1.480	1.545	1.605	1.660	1.715	1.760	1.810
13.5	1.090	1.175	1.255	1.330	1.405	1.470	1.530	1.590	1.645	1.700	1.745	1.790
14.0	1.090	1.175	1.250	1.325	1.395	1.460	1.520	1.580	1.635	1.685	1.730	1.775
14.5	1.085	1.170	1,245	1.315	1.385	1.445	1.505	1.565	1.620	1.670	1.715	1.760
15.0	1.085	1.165	1.235	1.310	1.355	1.440	1.500	1.555	1.610	1.655	1.705	1.750

Table of various powers of n for 5 decimal places

n	n^2	n^3	n^4	n	n^2	n^3	n^4	n	n^2	n^3	n^4
0.00	0.00000	0.00000	0.00000	0.34	0.11560	0.03930	0.01336	0.68	0.46240	0.31443	0.21381
0.01	.00010	.00000	.00000	0.35	.12250	.04288	.01501	0.69	.47610	.32851	.22667
0.02	.00040	.00001	.00000	0.36	.12960	.04666	.01680	0.70	.49000	.34300	.24010
0.03	.00090	.00003	.00000	0.37	.13690	.05065	.01874				
0.04	.00160	.00006	.00000	0.38	.14440	.05487	.02085	0.71	.50410	.35791	.25412
0.05	.00250	.00013	.00001	0.39	.15210	.05932	.02313	0.72	.51840	.37325	.26874
0.06	.00360	.00022	.00001					0.73	.52290	.38902	.28398
0.07	.00490	.00034	.00002	0.40	.16000	.06400	.02560	0.74	.54760	.40522	.29987
0.08	.00640	.00051	.00004					0.75	.56250	.42188	.31641
0.09	.00810	.00073	.00007	0.41	.16810	.06892	.02826	0.76	.57760	.43898	.33362
				0.42	.17640	.07409	.03112	0.77	.59290	.45653	.35153
0.10	.01000	.00100	.00010	0.43	.18490	.07951	.03419	0.78	.60840	.47455	.37015
				0.44	.19360	.08518	.03748	0.79	.62410	.49304	.38950
0.11	.01210	.00133	.00015	0.45	.20250	.09113	.04101				
0.12	.01440	.00173	.00021	0.46	.21160	.09734	.04478	0.80	.64000	.51200	.40960
0.13	.01690	.00220	.00029	0.47	.22090	.10382	.04880				
0.14	.01960	.00274	.00038	0.48	.23040	.11059	.05308	0.81	.65610	.53144	.43047
0.15	.02250	.00338	.00051	0.49	.24010	.11765	.05765	0.82	.67240	.55137	.45212
0.16	.02560	.00410	.00061					0.83	.68890	.57179	.47458
0.17	.02890	.00491	.00084	0.50	.25000	.12500	.06250	0.84	.70560	.59270	.49787
0.18	.03240	.00583	.00105					0.85	.72250	.61413	.52201
0.19	.03610	.00686	.00130	0.51	.26010	.13265	.06765	0.86	.73960	.63606	.54701
				0.52	.27040	.14061	.07312	0.87	.75690	.65850	.57290
0.20	.04000	.00800	.00160	0.53	.28090	.14888	.07890	0.88	.77440	.68147	.59970
				0.54	.29160	.15746	.08503	0.89	.79210	.70497	.62742
0.21	.04410	.00926	.00195	0.55	.30250	.16638	.09151				
0.22	.04840	.01065	.00234	0.56	.31360	.17562	.09835	0.90	.81000	.72900	.65610
0.23	.05290	.01217	.00280	0.57	.32490	.18519	.10556				
0.24	.05760	.01382	.00332	0.58	.33640	.19511	.11317	0.91	.82810	.75357	.68575
0.25	.06250	.01563	.00391	0.59	.34810	.20538	.12117	0.92	.84640	.77869	.71639
0.26	.06760	.01758	.00457					0.93	.86490	.80436	.74805
0.27	.07290	.01968	.00531	0.60	.36000	.21600	.12960	0.94	.88360	.83058	.78075
0.28	.07840	.02195	.00615					0.95	.90250	.85738	.81451
0.29	.08410	.02439	.00707	0.61	.37210	.22698	.13846	0.96	.92160	.88474	.84935
				0.62	.38440	.23833	.14776	0.97	.94090	.91267	.88529
0.30	.09000	.02700	.00810	0.63	.39690	.25005	.15753	0.98	.96040	.94119	.92237
				0.64	.40960	.26214	.16777	0.99	.98010	.97030	.96060
0.31	.09610	.02979	.00924	0.65	.42250	.27463	.17851	1.00	1.00000	1.00000	1.00000
0.32	.10240	.03277	.01049	0.66	.43560	.28750	.18975				
0.33	.10890	.03594	.01186	0.67	.44890	.30076	.20151				

Loading 1 (to Table 1)

Bound line load, uniformly distributed

$$s = n\,l \qquad\qquad 0 \leq n \leq 1$$

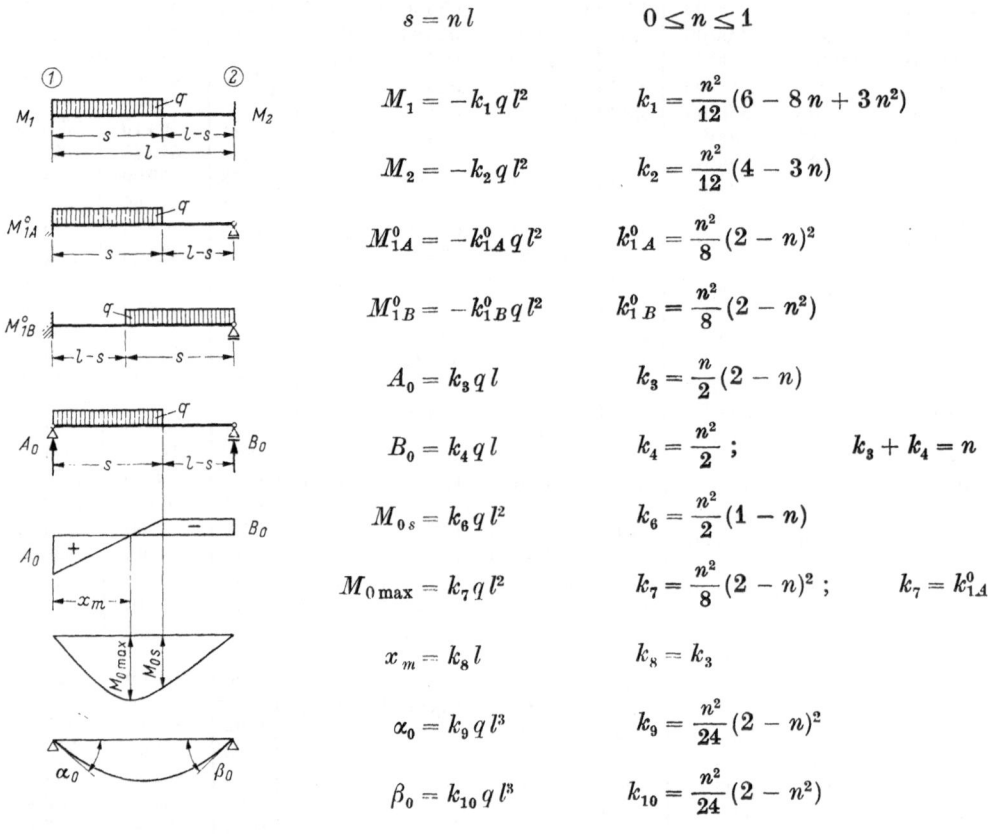

$$M_1 = -k_1\,q\,l^2 \qquad k_1 = \frac{n^2}{12}\,(6 - 8\,n + 3\,n^2)$$

$$M_2 = -k_2\,q\,l^2 \qquad k_2 = \frac{n^2}{12}\,(4 - 3\,n)$$

$$M_{1A}^0 = -k_{1A}^0\,q\,l^2 \qquad k_{1A}^0 = \frac{n^2}{8}\,(2 - n)^2$$

$$M_{1B}^0 = -k_{1B}^0\,q\,l^2 \qquad k_{1B}^0 = \frac{n^2}{8}\,(2 - n^2)$$

$$A_0 = k_3\,q\,l \qquad k_3 = \frac{n}{2}\,(2 - n)$$

$$B_0 = k_4\,q\,l \qquad k_4 = \frac{n^2}{2}\,; \qquad k_3 + k_4 = n$$

$$M_{0s} = k_6\,q\,l^2 \qquad k_6 = \frac{n^2}{2}\,(1 - n)$$

$$M_{0\,max} = k_7\,q\,l^2 \qquad k_7 = \frac{n^2}{8}\,(2 - n)^2\,; \qquad k_7 = k_{1A}^0$$

$$x_m = k_8\,l \qquad k_8 = k_3$$

$$\alpha_0 = k_9\,q\,l^3 \qquad k_9 = \frac{n^2}{24}\,(2 - n)^2$$

$$\beta_0 = k_{10}\,q\,l^3 \qquad k_{10} = \frac{n^2}{24}\,(2 - n^2)$$

Table 1

Table 1

n	k_1	k_2	k^0_{1A}	k^0_{1B}	k_3	k_4	k_6	k_7	k_9	k_{10}	n
0	0	0	0	0	0	0	0	0	0	0	0
0.05	0.001	0.000	0.001	0.000	0.049	0.001	0.002	0.001	0.000	0.000	0.05
0.1	.004	.000	.005	.003	.095	.005	.005	.005	.002	.001	0.1
0.15	.009	.001	.010	.006	.139	.011	.010	.010	.003	.002	0.15
0.2	.015	.002	.016	.010	.180	.020	.016	.016	.005	.003	0.2
0.25	.022	.004	.024	.015	.219	.031	.023	.024	.008	.005	0.25
0.3	.029	.007	.033	.022	.255	.045	.032	.033	.011	.007	0.3
0.35	.036	.011	.042	.029	.289	.061	.040	.042	.014	.010	0.35
0.4	.044	.015	.051	.037	.320	.080	.048	.051	.017	.012	0.4
0.45	.051	.020	.061	.046	.349	.101	.056	.061	.020	.015	0.45
0.5	.057	.026	.070	.055	.375	.125	.063	.070	.023	.018	0.5
0.55	.063	.033	.080	.064	.399	.151	.068	.080	.027	.021	0.55
0.6	.068	.040	.088	.074	.420	.180	.072	.088	.029	.025	0.6
0.65	.073	.047	.096	.083	.439	.211	.074	.096	.032	.028	0.65
0.7	.076	.054	.104	.093	.455	.245	.074	.104	.035	.031	0.7
0.75	.079	.062	.110	.101	.469	.281	.070	.110	.037	.034	0.75
0.8	.081	.068	.115	.109	.480	.320	.064	.115	.038	.036	0.8
0.85	.082	.074	.119	.115	.489	.361	.054	.119	.040	.039	0.85
0.9	.083	.079	.123	.121	.495	.405	.041	.123	.041	.040	0.9
0.95	.083	.082	.124	.124	.499	.451	.023	.124	.042	.041	0.95
1.0	0.083	0.083	0.125	0.125	0.500	0.500	0.000	0.125	0.042	0.042	1.0

Table 2   – 30 –

Loading 2 (to Table 2)

Symmetrical line load, uniformly distributed

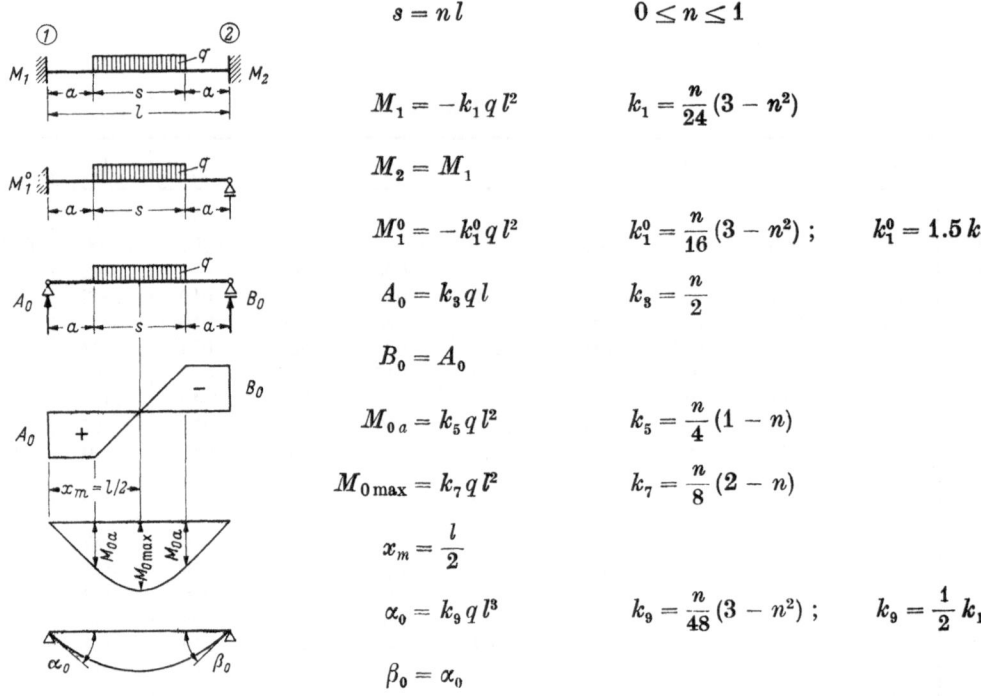

$$s = nl \qquad\qquad 0 \le n \le 1$$

$$M_1 = -k_1 q l^2 \qquad k_1 = \frac{n}{24}(3 - n^2)$$

$$M_2 = M_1$$

$$M_1^0 = -k_1^0 q l^2 \qquad k_1^0 = \frac{n}{16}(3 - n^2) \;; \qquad k_1^0 = 1.5\,k_1$$

$$A_0 = k_3 q l \qquad k_3 = \frac{n}{2}$$

$$B_0 = A_0$$

$$M_{0a} = k_5 q l^2 \qquad k_5 = \frac{n}{4}(1 - n)$$

$$M_{0\,\text{max}} = k_7 q l^2 \qquad k_7 = \frac{n}{8}(2 - n)$$

$$x_m = \frac{l}{2}$$

$$\alpha_0 = k_9 q l^3 \qquad k_9 = \frac{n}{48}(3 - n^2) \;; \qquad k_9 = \frac{1}{2}\,k_1$$

$$\beta_0 = \alpha_0$$

Table 2

n	k_1	k_1^0	k_3	k_5	k_7	k_9	n
0	0	0	0	0	0	0	0
0.05	0.006	0.009	0.025	0.012	0.012	0.003	0.05
0.1	.013	.019	.050	.023	.024	.006	0.1
0.15	.019	.028	.075	.032	.035	.009	0.15
0.2	.025	.037	.100	.040	.045	.012	0.2
0.25	.031	.046	.125	.047	.055	.015	0.25
0.3	.036	.055	.150	.053	.064	.018	0.3
0.35	.042	.063	.175	.057	.072	.021	0.35
0.4	.047	.071	.200	.060	.080	.024	0.4
0.45	.053	.079	.225	.062	.087	.026	0.45
0.5	.057	.086	.250	.063	.094	.029	0.5
0.55	.062	.093	.275	.062	.100	.031	0.55
0.6	.066	.099	.300	.060	.105	.033	0.6
0.65	.070	.105	.325	.057	.110	.035	0.65
0.7	.073	.110	.350	.053	.114	.037	0.7
0.75	.076	.114	.375	.047	.117	.038	0.75
0.8	.079	.118	.400	.040	.120	.039	0.8
0.85	.081	.121	.425	.032	.122	.040	0.85
0.9	.082	.123	.450	.023	.124	.041	0.9
0.95	.083	.125	.475	.012	.125	.042	0.95
1.0	0.083	0.125	0.500	0.000	0.125	0.042	1.0

Table 3

Loading 3 (to Table 3)

Symmetrical twin line load, uniformly distributed

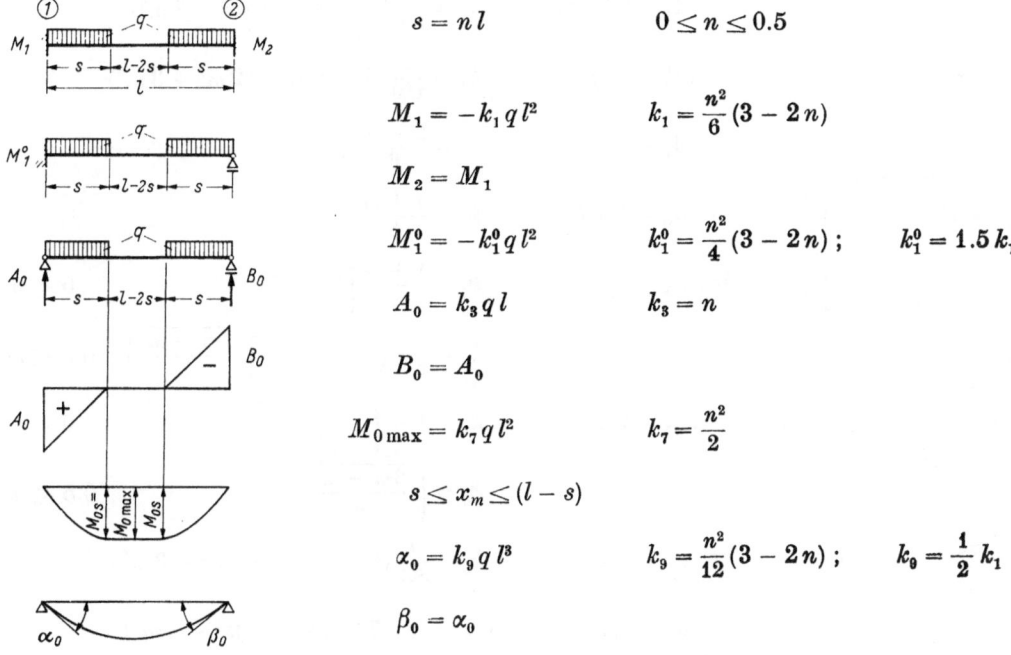

$$s = nl \qquad 0 \leq n \leq 0.5$$

$$M_1 = -k_1 q l^2 \qquad k_1 = \frac{n^2}{6}(3 - 2n)$$

$$M_2 = M_1$$

$$M_1^0 = -k_1^0 q l^2 \qquad k_1^0 = \frac{n^2}{4}(3 - 2n); \qquad k_1^0 = 1.5\,k_1$$

$$A_0 = k_3 q l \qquad k_3 = n$$

$$B_0 = A_0$$

$$M_{0\,max} = k_7 q l^2 \qquad k_7 = \frac{n^2}{2}$$

$$s \leq x_m \leq (l - s)$$

$$\alpha_0 = k_9 q l^3 \qquad k_9 = \frac{n^2}{12}(3 - 2n); \qquad k_9 = \frac{1}{2}k_1$$

$$\beta_0 = \alpha_0$$

Table 3

n	k_1	k_1^0	k_3	k_7	k_9	n
0	0	0	0	0	0	0
0.05	0.001	0.002	0.050	0.001	0.001	0.05
0.1	.005	.007	.100	.005	.002	0.1
0.15	.010	.015	.150	.011	.005	0.15
0.2	.017	.026	.200	.020	.009	0.2
0.25	.026	.039	.250	.031	.013	0.25
0.3	.036	.054	.300	.045	.018	0.3
0.35	.047	.070	.350	.061	.024	0.35
0.4	.059	.088	.400	.080	.029	0.4
0.45	.071	.106	.450	.101	.035	0.45
0.5	0.083	0.125	0.500	0.125	0.042	0.5

Table 4   – 32 –

Loading 4 (to Table 4)

Continuous triangular load

$$a = m\,l \qquad\qquad 0 \le m \le 1$$

$$M_1 = -k_1\,q\,l^2 \qquad k_1 = \frac{1}{60}(3 + 3m - 7m^2 + 3m^3)$$

$$M_2 = -k_2\,q\,l^2 \qquad k_2 = \frac{1}{60}(2 + 2m + 2m^2 - 3m^3)$$

$$M_1^0 = k_1^0\,q\,l^2 \qquad k_1^0 = \frac{1}{120}(8 + 8m - 12m^2 + 3m^3)$$

$$A_0 = k_3\,q\,l \qquad k_3 = \frac{2-m}{6}$$

$$B_0 = k_4\,q\,l \qquad k_4 = \frac{1+m}{6}$$

$$M_{0\,max} = k_7\,q\,l_2 \qquad k_7 = \frac{1+m}{9}\sqrt{\frac{1-m^2}{3}} \qquad 0 \le m \le 0.5$$

$$k_7 = \frac{1}{6m}\left(\frac{4m-2m^2}{3}\sqrt{\frac{2m-m^2}{3}}\right) \qquad 0.5 \le m \le 1.0$$

$$x_m = k_8\,l \qquad k_8 = 1 - \sqrt{\frac{1-m^2}{3}} \qquad 0 \le m \le 0.5$$

$$k_8 = \sqrt{\frac{2m-m^2}{3}} \qquad 0.5 \le m \le 1.0$$

$$\alpha_0 = k_9\,q\,l^3 \qquad k_9 = \frac{1}{360}(8 + 8m - 12m^2 + 3m^3)$$

$$\beta_0 = k_{10}\,q\,l^3 \qquad k_{10} = \frac{1}{360}(7 + 7m - 3m^2 - 3m^3)$$

Table 4

m	k_1	k_2	k_1^0	k_3	k_4	k_7	k_8	k_9	k_{10}	m
0	0.050	0.033	0.067	0.333	0.167	0.064	0.423	0.022	0.019	0
0.05	.052	.035	.070	.325	.175	.067	.423	.023	.020	0.05
0.1	.055	.037	.072	.317	.183	.070	.426	.024	.021	0.1
0.15	.055	.039	.075	.308	.192	.073	.429	.025	.021	0.15
0.2	.056	.041	.076	.300	.200	.075	.434	.025	.023	0.2
0.25	.056	.043	.078	.292	.208	.078	.441	.026	.024	0.25
0.3	.056	.045	.078	.283	.217	.080	.449	.026	.024	0.3
0.35	.055	.047	.079	.275	.225	.081	.459	.026	.025	0.35
0.4	.055	.049	.079	.267	.233	.082	.471	.026	.025	0.4
0.45	.053	.051	.079	.258	.242	.083	.484	.026	.026	0.45
0.5	.052	.052	.078	.250	.250	.083	.500	.026	.026	0.5
0.55	.051	.053	.077	.242	.258	.083	.516	.026	.026	0.55
0.6	.049	.055	.076	.233	.267	.082	.529	.025	.026	0.6
0.65	047	.055	.075	.225	.275	.081	.541	.025	.026	0.65
0.7	.045	.056	.073	.217	.283	.080	.551	.024	.026	0.7
0.75	.043	.056	.071	.208	.292	.078	.559	.024	.026	0.75
0.8	.041	.056	.069	.200	.300	.075	.566	.023	.025	0.8
0.85	.039	.055	.066	.192	.308	.073	.571	.022	.025	0.85
0.9	.037	.054	.064	.183	.317	.070	.574	.021	.024	0.9
0.95	.035	.052	.061	.175	.325	.067	.577	.020	.023	0.95
1.0	0.033	0.050	0.058	0.167	0.333	0.064	0·577	0.019	0.022	1.0

Loading 5 (to Table 5)

Bound triangular line load

$$s = n\,l \qquad\qquad 0 \leq n \leq 1.0$$

$$M_1 = -k_1\,q\,l^2 \qquad\qquad k_1 = \frac{n^2}{60}(10 - 10\,n + 3\,n^2)$$

$$M_2 = -k_2\,q\,l^2 \qquad\qquad k_2 = \frac{n^3}{60}(5 - 3\,n)$$

$$M_{1A}^0 = -k_{1A}^0\,q\,l^2 \qquad\qquad k_{1A}^0 = \frac{n^2}{120}(20 - 15\,n + 3\,n^2)$$

$$M_{1B}^0 = -k_{1B}^0\,q\,l^2 \qquad\qquad k_{1B}^0 = \frac{n^2}{120}(10 - 3\,n^2)$$

$$A_0 = k_3\,q\,l \qquad\qquad k_3 = \frac{n}{6}(3 - n)$$

$$B_0 = k_4\,q\,l \qquad\qquad k_4 = \frac{n^2}{6}$$

$$M_{0s} = k_6\,q\,l^2 \qquad\qquad k_6 = \frac{n^2}{6}(1 - n)$$

$$M_{0\max} = k_7\,q\,l^2 \qquad\qquad k_7 = \frac{n^2}{6}\left(1 - n + \frac{2\,n}{3}\sqrt{\frac{n}{3}}\right)$$

$$x_m = k_8\,l \qquad\qquad k_8 = n\left(1 - \sqrt{\frac{n}{3}}\right)$$

$$\alpha_0 = k_9\,q\,l^3 \qquad\qquad k_9 = \frac{n^2}{360}(20 - 15\,n + 3\,n^2)$$

$$\beta_0 = k_{10}\,q\,l^3 \qquad\qquad k_{10} = \frac{n^2}{360}(10 - 3\,n^2)$$

Table 5

n	k_1	k_2	k_{1A}^0	k_{1B}^0	k_3	k_4	k_6	k_7	k_8	k_9	k_{10}	n
0	0	0	0	0	0	0	0	0	0	0	0	0
0.05	0.000	0.000	0.000	0.000	0.025	0.001	0.000	0.000	0.044	0.000	0.000	0.005
0.1	.002	.000	.002	.001	.048	.002	.002	.002	.082	.001	.000	0.1
0.15	.003	.000	.003	.002	.071	.004	.003	.003	.116	.001	.001	0.15
0.2	.005	.001	.006	.003	.093	.007	.005	.006	.148	.002	.001	0.02
0.25	.008	.001	.009	.005	.115	.010	.008	.008	.178	.003	.02	0.025
0.3	.011	.002	.012	.007	.135	.015	.011	.011	.205	.004	.002	0.3
0.35	.014	.003	.015	.010	.155	.020	.013	.015	.230	.005	.003	0.35
0.4	.017	.004	.019	.013	.173	.027	.016	.019	.254	.006	.004	0.4
0.45	.021	.006	.023	.016	.191	.034	.019	.023	.276	.008	.005	0.45
0.5	.024	.007	.028	.019	.208	.042	.021	.027	.296	.009	.006	0.5
0.55	.027	.009	.032	.023	.225	.050	.023	.031	.315	.011	.008	0.55
0.6	.030	.012	.036	.027	.240	.060	.024	.035	.332	.012	.009	0.6
0.65	.034	.014	.041	.031	.255	.070	.025	.039	.347	.014	.010	0.65
0.7	.037	.017	.045	.035	.268	.082	.025	.043	.362	.015	.012	0.7
0.75	.039	.019	.049	.039	.281	.094	.023	.047	.375	.016	.013	0.75
0.8	.042	.022	.053	.043	.293	.107	.021	.051	.387	.018	.014	0.8
0.85	.044	.025	.057	.047	.305	.120	.018	.054	.398	.019	.016	0.85
0.9	.046	.028	.060	.051	.315	.135	.014	.058	.407	.020	.017	0.9
0.95	.048	.031	.064	.055	.325	.150	.008	.061	.415	.021	.018	0.95
1.0	0.050	0.033	0.067	0.058	0.333	0.167	0.000	0.064	0.423	0.022	0.019	1.0

Table 6   – 34 –

Loading 6 (to Table 6)

Symmetrically bound triangular twin line load

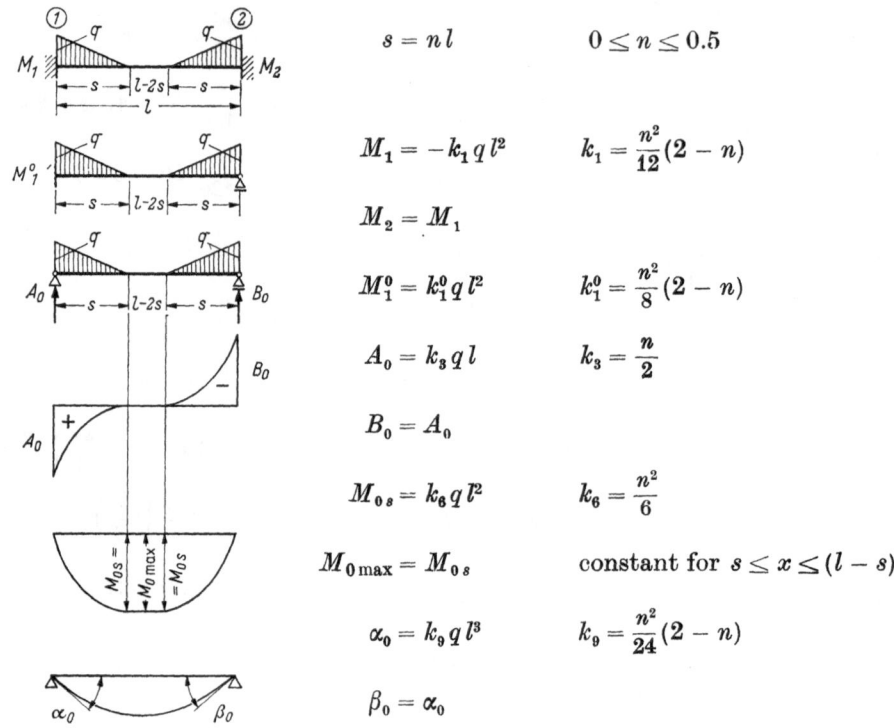

$$s = n\,l \qquad\qquad 0 \leq n \leq 0.5$$

$$M_1 = -k_1\,q\,l^2 \qquad k_1 = \frac{n^2}{12}(2-n)$$

$$M_2 = M_1$$

$$M_1^0 = k_1^0\,q\,l^2 \qquad k_1^0 = \frac{n^2}{8}(2-n)$$

$$A_0 = k_3\,q\,l \qquad\quad k_3 = \frac{n}{2}$$

$$B_0 = A_0$$

$$M_{0\,s} = k_6\,q\,l^2 \qquad\quad k_6 = \frac{n^2}{6}$$

$$M_{0\,\max} = M_{0\,s} \qquad \text{constant for } s \leq x \leq (l-s)$$

$$\alpha_0 = k_9\,q\,l^3 \qquad\quad k_9 = \frac{n^2}{24}(2-n)$$

$$\beta_0 = \alpha_0$$

Table 6

n	k_1	k_1^0	k_3	k_6	k_9	n
0	0	0	0	0	0	0
0.05	0.000	0.001	0.025	0.000	0.000	0.05
0.1	.002	.002	.050	.002	.001	0.1
0.15	.003	.005	.075	.004	.002	0.15
0.2	.006	.009	.100	.007	.003	0.2
0.25	.009	.014	.125	.010	.005	0.25
0.3	.013	.019	.150	.015	.006	0.3
0.35	.017	.025	.175	.020	.008	0.35
0.4	.021	.032	.200	.027	.011	0.4
0.45	.026	.039	.225	.034	.013	0.45
0.5	0.031	0.047	0.250	0.042	0.016	0.5

Loading 7 (to Table 7)

Symmetrical twin triangular load

$$s = n\,l \qquad\qquad 0 \leq n \leq 0.5$$

$$M_1 = -k_1\,q\,l^2 \qquad k_1 = \frac{n}{8}(1 - 2\,n^2)$$

$$M_2 = M_1$$

$$M_1^0 = -k_1^0\,q\,l^2 \qquad k_1^0 = \frac{n}{16}(3 - 6\,n^2)\,; \qquad k_1^0 = 1.5\,k_1$$

$$A_0 = k_3\,q\,l \qquad k_3 = \frac{n}{2}$$

$$B_0 = A_0$$

$$M_{0\,s} = k_5\,q\,l^2 \qquad k_5 = \frac{n}{4}(1 - 2\,n)$$

$$M_{0\,\max} = k_7\,q\,l^2 \qquad k_7 = \frac{n}{12}(3 - 4\,n)$$

$$x_m = \frac{l}{2}$$

$$\alpha_0 = k_9\,x\,q\,l^3 \qquad k_9 = \frac{n}{16}(1 - 2\,n^2)\,; \qquad k_9 - \frac{1}{2}\,k_1$$

$$\beta_0 = \alpha_0$$

Table 7

n	k_1	k_1^0	k_3	k_5	k_7	k_9	n
0	0	0	0	0	0	0	0
0.05	0.006	0.009	0.025	0.011	0.012	0.003	0.05
0.1	.012	.018	.050	.020	.022	.006	0.1
0.15	.018	.027	.075	.026	.030	.009	0.15
0.2	.023	.035	.100	.030	.037	.012	0.2
0.25	.027	.041	.125	.031	.042	.014	0.25
0.3	.031	.046	.150	.030	.045	.015	0.3
0.35	.033	.050	.175	.026	.047	.017	0.35
0.4	.034	.051	.200	.020	.047	.017	0.4
0.45	.033	.050	.225	.011	.045	.017	0.45
0.5	0.031	0.047	0.250	0.00	0.042	0.016	0.5

Table 8   – 36 –

Loading 8 (to Table 8)

Bound triangular line load

$$s = n l \qquad\qquad 0 \leq n \leq 1$$

$$M_1 = -k_1 q l^2 \qquad k_1 = \frac{n^2}{30}(10 - 15n + 6n^2)$$

$$M_2 = -k_2 q l^2 \qquad k_2 = \frac{n^2}{20}(5 - 4n)$$

$$M_{1A}^0 = -k_{1A}^0 q l^2 \qquad k_{1A}^0 = \frac{n^2}{120}(40 - 45n + 12n^2)$$

$$M_{1B}^0 = -k_{1B}^0 q l^2 \qquad k_{1B}^0 = \frac{n^2}{30}(5 - 3n^2)$$

$$A_0 = k_3 q l \qquad k_3 = \frac{n}{6}(3 - 2n)$$

$$B_0 = k_4 q l \qquad k_4 = \frac{n^2}{3}$$

$$M_{0s} = k_6 q l^2 \qquad k_6 = \frac{n^2}{3}(1 - n)$$

$$M_{0\max} = k_7 q l^2 \qquad k_7 = \frac{n^2}{3}\sqrt{\left(\frac{3 - 2n}{3}\right)^3}$$

$$x_m = k_8 l \qquad k_8 = n\sqrt{\frac{3 - 2n}{3}}$$

$$\alpha_0 = k_9 q l^3 \qquad k_9 = \frac{n^2}{360}(40 - 45n + 12n^2)$$

$$\beta_0 = k_{10} q l^3 \qquad k_{10} = \frac{n^2}{90}(5 - 3n^2)$$

Table 8

n	k_1	k_2	k_{1A}^0	k_{1B}^0	k_3	k_4	k_6	k_7	k_8	k_9	k_{10}	n
0	0	0	0	0	0	0	0	0	0	0	0	0
0.05	0.001	0.000	0.001	0.000	0.024	0.001	0.001	0.001	0.049	0.000	0.000	0.05
0.1	.003	.000	.003	.002	.047	.003	.003	.003	.097	.001	.001	0.1
0.15	.006	.001	.006	.004	.068	.008	.006	.006	.142	.002	.001	0.15
0.2	.010	.002	.011	.007	.087	.013	.011	.011	.186	.004	.002	0.2
0.25	.014	.003	.015	.010	.104	.021	.016	.016	.228	.005	.003	0.25
0.3	.018	.005	.021	.014	.120	.030	.021	.022	.268	.007	.005	0.3
0.35	.022	.008	.026	.019	.134	.041	.027	.027	.306	.009	.006	0.35
0.4	.026	.011	.032	.024	.147	.053	.032	.034	.343	.011	.008	0.4
0.45	.030	.015	.037	.030	.158	.068	.037	.040	.377	.012	.010	0.45
0.5	.033	.019	.043	.035	.167	.083	.042	.045	.409	.014	.012	0.5
0.55	.036	.023	.048	.041	.174	.101	.045	.051	.438	.016	.014	0.55
0.6	.038	.028	.052	.047	.180	.120	.048	.056	.465	.017	.016	0.6
0.65	.039	.033	.056	.053	.184	.141	.049	.060	.489	.019	.018	0.65
0.7	.040	.038	.059	.058	.187	.163	.049	.064	.511	.020	.019	0.7
0.75	.040	.042	.061	.062	.188	.188	.047	.066	.530	.020	.021	0.75
0.8	.039	.046	.062	.066	.187	.213	.043	.068	.547	.021	.022	0.8
0.85	.038	.049	.063	.068	.184	.241	.036	.069	.555	.021	.023	0.85
0.9	.037	.051	.062	.069	.180	.270	.027	.068	.569	.021	.023	0.9
0.95	.035	.051	.061	.069	.174	.301	.015	.067	.575	.020	.023	0.95
1.0	0.033	0.050	0.058	0.067	0.167	0.333	0.000	0.064	0.577	0.019	0.022	1.0

Table 9

Loading 9 (to Table 9)

Symmetrically bound triangular load

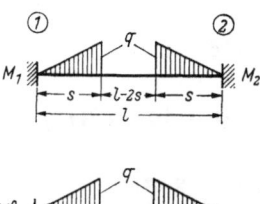

$$s = n\,l \qquad\qquad 0 \leq n \leq 0.5$$

$$M_1 = -k_1\,q\,l^2 \qquad k_1 = \frac{n^2}{12}\,(4 - 3\,n)$$

$$M_2 = M_1$$

$$M_1^0 = -k_1^0\,q\,l^2 \qquad k_1^0 = \frac{n^2}{8}\,(4 - 3\,n)\,; \qquad k_1^0 = 1.5\,k_1$$

$$A_0 = k_3\,q\,l \qquad k_3 = \frac{n}{2}$$

$$B_0 = A_0$$

$$M_{0s} = k_6\,q\,l^2 \qquad k_6 = \frac{n^2}{3}$$

$$M_{0\,max} = M_{0s} \qquad \text{constant for } s \leq x \leq (l - s)$$

$$\alpha_0 = k_9\,q\,l^3 \qquad k_9 = \frac{n^2}{24}\,(4 - 3\,n)$$

$$\beta_0 = \alpha_0$$

Table 9

n	k_1	k_1^0	k_3	k_6	k_9	n
0	0	0	0	0	0	0
0.05	0.001	0.001	0.025	0.001	0.000	0.005
0.1	.003	.005	.050	.003	.002	0.1
0.15	.007	.010	.075	.005	.003	0.15
0.2	.011	.017	.100	.013	.006	0.2
0.25	.017	.025	.125	.021	.008	0.25
0.3	.023	.035	.150	.030	.012	0.3
0.35	.030	.045	.175	.041	.015	0.35
0.4	.037	.056	.200	.053	.019	0.4
0.45	.045	.067	.225	.068	.022	0.45
0.5	0.052	0.078	0.250	0.083	0.026	0.5

Table 10   – 38 –

Loading 10 (to Table 10)

Bound triangular line load

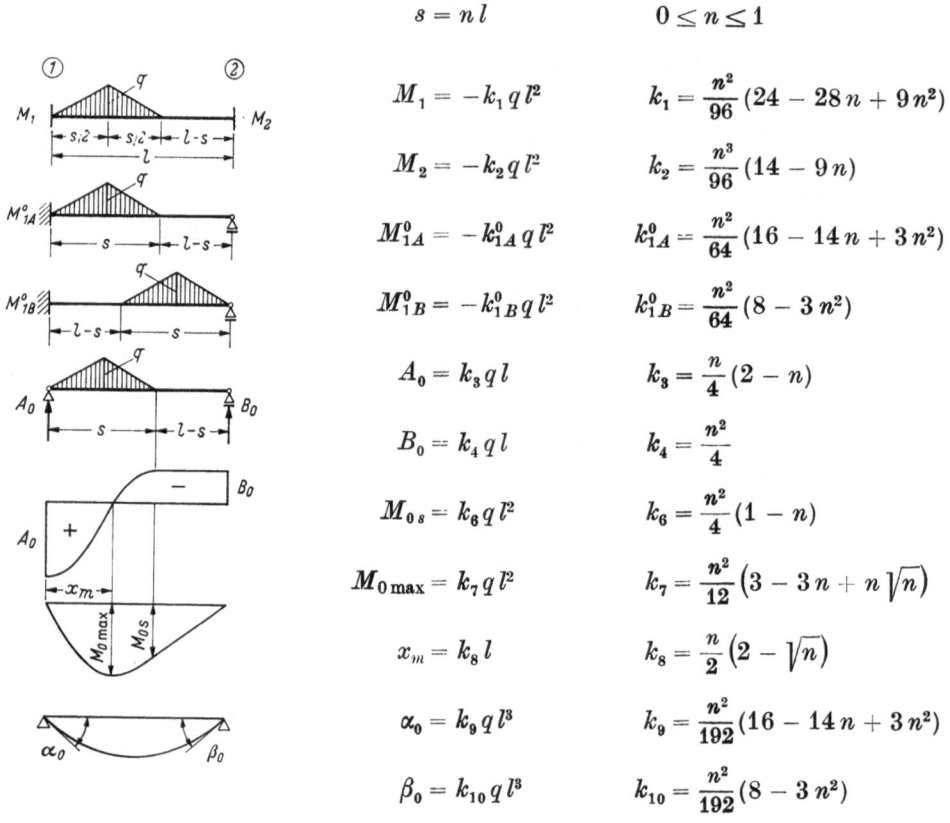

$$s = n\,l \qquad\qquad 0 \le n \le 1$$

$$M_1 = -k_1\,q\,l^2 \qquad k_1 = \frac{n^2}{96}\,(24 - 28\,n + 9\,n^2)$$

$$M_2 = -k_2\,q\,l^2 \qquad k_2 = \frac{n^3}{96}\,(14 - 9\,n)$$

$$M^0_{1A} = -k^0_{1A}\,q\,l^2 \qquad k^0_{1A} = \frac{n^2}{64}\,(16 - 14\,n + 3\,n^2)$$

$$M^0_{1B} = -k^0_{1B}\,q\,l^2 \qquad k^0_{1B} = \frac{n^2}{64}\,(8 - 3\,n^2)$$

$$A_0 = k_3\,q\,l \qquad k_3 = \frac{n}{4}\,(2 - n)$$

$$B_0 = k_4\,q\,l \qquad k_4 = \frac{n^2}{4}$$

$$M_{0s} = k_6\,q\,l^2 \qquad k_6 = \frac{n^2}{4}\,(1 - n)$$

$$M_{0\,max} = k_7\,q\,l^2 \qquad k_7 = \frac{n^2}{12}\,(3 - 3\,n + n\,\sqrt{n})$$

$$x_m = k_8\,l \qquad k_8 = \frac{n}{2}\,(2 - \sqrt{n})$$

$$\alpha_0 = k_9\,q\,l^3 \qquad k_9 = \frac{n^2}{192}\,(16 - 14\,n + 3\,n^2)$$

$$\beta_0 = k_{10}\,q\,l^3 \qquad k_{10} = \frac{n^2}{192}\,(8 - 3\,n^2)$$

Table 10

n	k_1	k_2	k^0_{1A}	k^0_{1B}	k_3	k_4	k_6	k_7	k_8	k_9	k_{10}	n
0	0	0	0	0	0	0	0	0	0	0	0	0
0.05	0.001	0.000	0.001	0.000	0.024	0.001	0.001	0.001	0.044	0.000	0.000	0.05
0.1	.002	.000	.002	.001	.048	.003	.002	.002	.084	.001	.000	0.1
0.15	.005	.000	.005	.003	.069	.006	.005	.005	.121	.002	.001	0.15
0.2	.008	.001	.008	.005	.090	.010	.008	.008	.155	.003	.002	0.2
0.25	.011	.002	.012	.008	.109	.016	.012	.012	.188	.004	.003	0.25
0.3	.015	.003	.017	.011	.128	.023	.016	.017	.218	.006	.004	0.3
0.35	.020	.005	.022	.015	.144	.031	.020	.022	.246	.007	.005	0.35
0.4	.024	.007	.027	.019	.160	.040	.024	.027	.274	.009	.006	0.4
0.45	.028	.010	.033	.023	.174	.051	.028	.033	.299	.011	.008	0.45
0.5	.032	.012	.038	.028	.188	.063	.031	.039	.323	.013	.009	0.5
0.55	.036	.016	.044	.034	.199	.076	.034	.044	.346	.015	.011	0.55
0.6	.039	.019	.049	.039	.210	.090	.036	.050	.368	.016	.013	0.6
0.65	.042	.023	.054	.044	.219	.106	.037	.055	.388	.018	.015	0.65
0.7	.045	.028	.059	.050	.228	.123	.037	.061	.407	.020	.017	0.7
0.75	.047	.032	.063	.055	.234	.141	.035	.066	.425	.021	.019	0.75
0.8	.049	.036	.067	.061	.240	.160	.032	.070	.442	.022	.020	0.8
0.85	.050	.041	.071	.066	.244	.181	.027	.074	.458	.024	.022	0.85
0.9	.051	.045	.074	.071	.248	.203	.020	.078	.473	.025	.024	0.9
0.95	.052	.049	.076	.075	.249	.226	.011	.081	.487	.025	.025	0.95
1.0	0.052	0.052	0.078	0.078	0.250	0.250	0.000	0.083	0.500	0.026	0.026	1.0

Table 11

Loading 11 (to Table 11)

Symmetrical triangular load

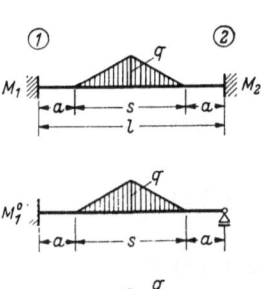

$$s = nl \qquad\qquad 0 \leq n \leq 1$$

$$M_1 = -k_1 q l^2 \qquad k_1 = \frac{n}{96}(6 - n^2)$$

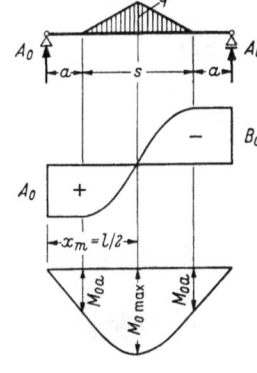

$$M_2 = M_1$$

$$M_1^0 = -k_1^0 q l^2 \qquad k_1^0 = \frac{n}{64}(6 - n^2); \qquad k_1^0 = 1.5\, k_1$$

$$A_0 = k_3 q l \qquad k_3 = \frac{n}{4}$$

$$B_0 = A_0$$

$$M_{0s} = k_5 q l^2 \qquad k_5 = \frac{n}{8}(1 - n)$$

$$M_{0\max} = k_7 q l^2 \qquad k_7 = \frac{n}{24}(3 - n)$$

$$x_m = \frac{l}{2}$$

$$\alpha_0 = k_9 q l^3 \qquad k_9 = \frac{n}{192}(6 - n^2); \qquad k_9 = \frac{1}{2} k_1$$

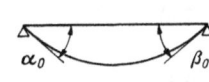

$$\beta_0 = \alpha_0$$

Table 11

n	k_1	k_1^0	k_3	k_5	k_7	k_9	n
0	0	0	0	0	0	0	0
0.05	0.003	0.005	0.013	0.006	0.006	0.002	0.05
0.1	.006	.009	.025	.011	.012	.003	0.1
0.15	.009	.014	.038	.016	.018	.005	0.15
0.2	.012	.019	.050	.020	.023	.006	0.2
0.25	.015	.023	.063	.023	.029	.008	0.25
0.3	.019	.028	.075	.026	.034	.009	0.3
0.35	.021	.032	.088	.028	.039	.011	0.35
0.4	.024	.037	.100	.030	.043	.012	0.4
0.45	.027	.041	.113	.031	.048	.014	0.45
0.5	.030	.045	.125	.031	.052	.015	0.5
0.55	.033	.049	.138	.031	.056	.016	0.55
0.6	.035	.053	.150	.030	.060	.018	0.6
0.65	.038	.057	.163	.028	.064	.019	0.65
0.7	.040	.060	.175	.026	.067	.020	0.7
0.75	.042	.064	.188	.023	.070	.021	0.75
0.8	.045	.067	.200	.020	.073	.022	0.8
0.85	.047	.070	.213	.016	.076	.023	0.85
0.9	.049	.073	.225	.011	.079	.024	0.9
0.95	.050	.076	.238	.006	.081	.025	0.95
1.0	0.052	0.078	0.250	0.000	0.083	0.026	1.0

Table 12   − 40 −

Loading 12 (to Table 12)

Continuous trapezoidal load

$$q' = \lambda q$$

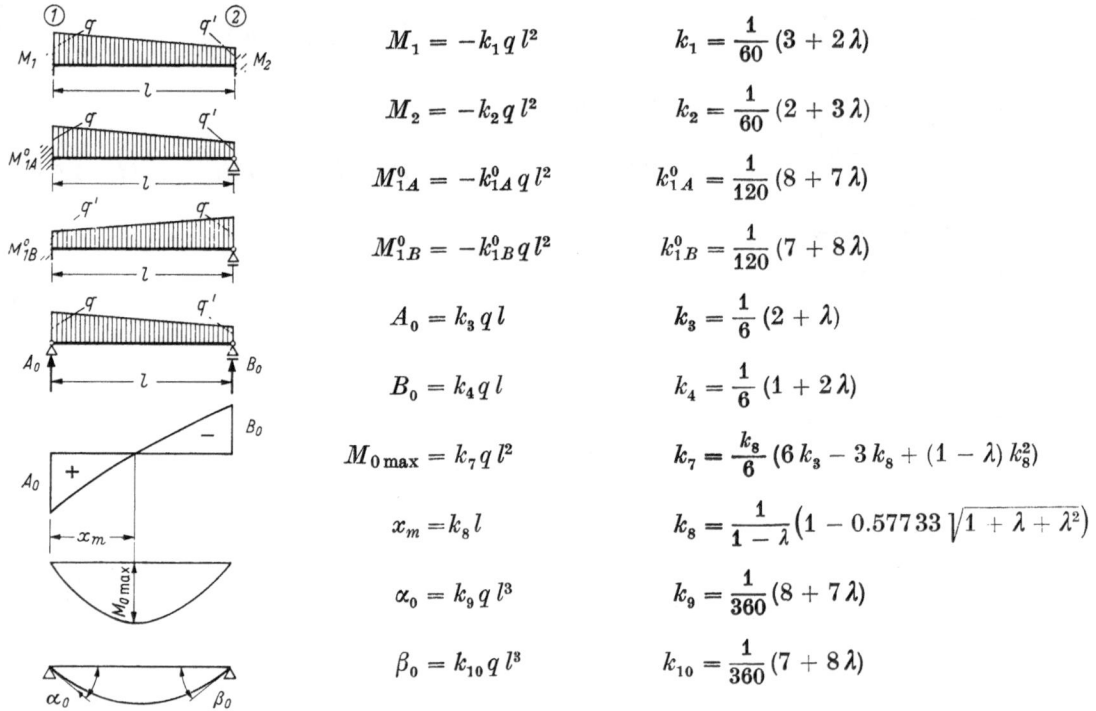

$$M_1 = -k_1 q l^2 \qquad k_1 = \frac{1}{60}(3 + 2\lambda)$$

$$M_2 = -k_2 q l^2 \qquad k_2 = \frac{1}{60}(2 + 3\lambda)$$

$$M_{1A}^0 = -k_{1A}^0 q l^2 \qquad k_{1A}^0 = \frac{1}{120}(8 + 7\lambda)$$

$$M_{1B}^0 = -k_{1B}^0 q l^2 \qquad k_{1B}^0 = \frac{1}{120}(7 + 8\lambda)$$

$$A_0 = k_3 q l \qquad k_3 = \frac{1}{6}(2 + \lambda)$$

$$B_0 = k_4 q l \qquad k_4 = \frac{1}{6}(1 + 2\lambda)$$

$$M_{0\,max} = k_7 q l^2 \qquad k_7 = \frac{k_8}{6}\left(6 k_3 - 3 k_8 + (1 - \lambda) k_8^2\right)$$

$$x_m = k_8 l \qquad k_8 = \frac{1}{1 - \lambda}\left(1 - 0.57733\sqrt{1 + \lambda + \lambda^2}\right)$$

$$\alpha_0 = k_9 q l^3 \qquad k_9 = \frac{1}{360}(8 + 7\lambda)$$

$$\beta_0 = k_{10} q l^3 \qquad k_{10} = \frac{1}{360}(7 + 8\lambda)$$

Table 12

λ	k_1	k_2	k_{1A}^0	k_{1B}^0	k_3	k_4	k_7	k_8	k_9	k_{10}	
0	0.050	0.033	0.067	0.058	0.333	0.167	0.064	0.423	0.022	0.019	0
0.05	.052	0.36	.070	.062	.342	.183	.067	.429	.023	.021	0.05
0.1	.053	.038	.073	.065	.350	.200	.070	.435	.024	.022	0.1
0.15	.055	.041	.075	.068	.358	.217	.073	.441	.025	.023	0.15
0.2	.057	.043	.078	.072	.367	.233	.076	.446	.026	.024	0.2
0.25	.058	.046	.081	.075	.375	.250	.079	.451	.027	.025	0.25
0.3	.060	.048	.084	.078	.383	.267	.082	.456	.028	.026	0.3
0.35	.062	.051	.087	.082	.392	.283	.085	.461	.029	.027	0.35
0.4	.063	.053	.090	.085	.400	.300	.088	.465	.030	.028	0.4
0.45	.065	.056	.093	.088	.408	.317	.091	.469	.031	.029	0.45
0.5	.067	.058	.096	.092	.417	.333	.094	.473	.032	.031	0.5
0.55	.068	.061	.099	.095	.425	.350	.097	.476	.033	.032	0.55
0.6	.070	.063	.102	.098	.433	.367	.100	.479	.034	.033	0.6
0.65	.072	.066	.105	.102	.442	.383	.103	.483	.035	.034	0.65
0.7	.073	.068	.108	.105	.450	.400	.106	.485	.036	.035	0.7
0.75	.075	.071	.110	.108	.458	.417	.109	.488	.037	.036	0.75
0.8	.077	.073	.113	.112	.467	.433	.113	.491	.038	.037	0.8
0.85	.078	.076	.116	.115	.475	.450	.116	.493	.039	.038	0.85
0.9	.080	.078	.119	.118	.483	.467	.119	.496	.040	.039	0.9
0.95	.082	.081	.122	.122	.492	.483	.122	.499	.041	.041	0.95
1.0	0.083	0.083	0.125	0.125	0.500	0.500	0.125	0.500	0.042	0.042	1.0

Loading 13 (to Table 13)

Bound trapezoidal load

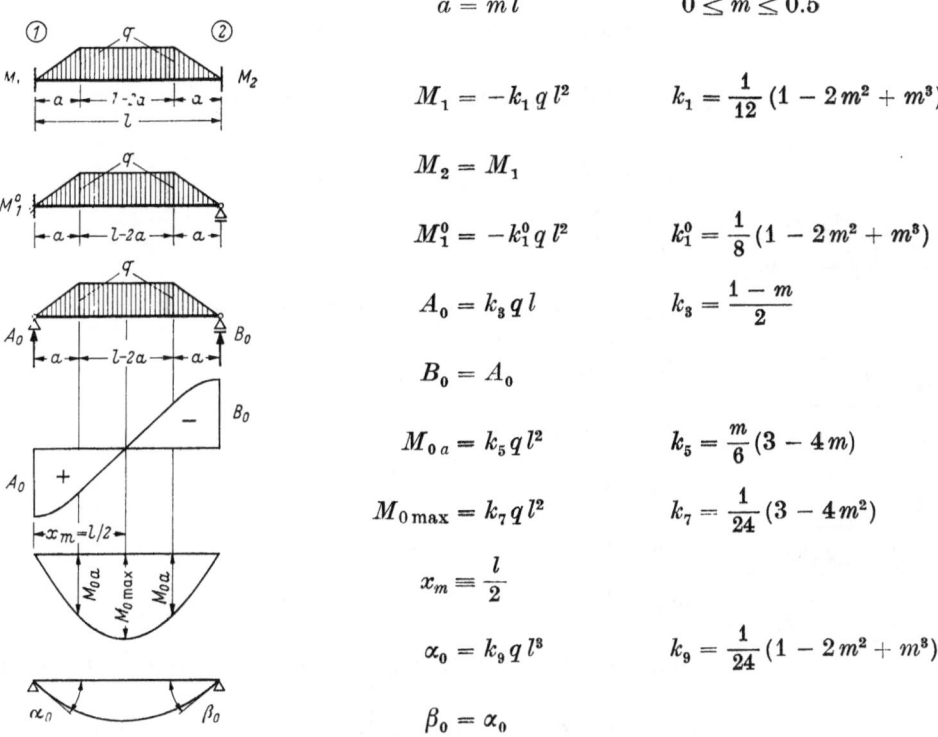

$$a = m\,l \qquad\qquad 0 \le m \le 0.5$$

$$M_1 = -k_1\,q\,l^2 \qquad k_1 = \frac{1}{12}(1 - 2\,m^2 + m^3)$$

$$M_2 = M_1$$

$$M_1^0 = -k_1^0\,q\,l^2 \qquad k_1^0 = \frac{1}{8}(1 - 2\,m^2 + m^3)$$

$$A_0 = k_3\,q\,l \qquad k_3 = \frac{1-m}{2}$$

$$B_0 = A_0$$

$$M_{0\,a} = k_5\,q\,l^2 \qquad k_5 = \frac{m}{6}(3 - 4\,m)$$

$$M_{0\,\max} = k_7\,q\,l^2 \qquad k_7 = \frac{1}{24}(3 - 4\,m^2)$$

$$x_m = \frac{l}{2}$$

$$\alpha_0 = k_9\,q\,l^3 \qquad k_9 = \frac{1}{24}(1 - 2\,m^2 + m^3)$$

$$\beta_0 = \alpha_0$$

Table 13

m	k_1	k_1^0	k_3	k_5	k_7	k_9	m
0	0.083	0.125	0.500	0.000	0.125	0.042	0
0.05	.083	.124	.475	.023	.125	.041	0.05
0.1	.082	.123	.450	.043	.123	.041	0.1
0.15	.080	.120	.425	.060	.121	.040	0.15
0.2	.077	.116	.400	.073	.118	.039	0.2
0.25	.074	.111	.375	.083	.115	.037	0.25
0.3	.071	.106	.350	.090	.110	.035	0.3
0.35	.067	.100	.325	.093	.105	.033	0.35
0.4	.062	.093	.300	.093	.098	.031	0.4
0.45	.057	.086	.275	.090	.091	.029	0.45
0.5	0.052	0.078	0.250	0.083	0.083	0.026	0.5

Loading 14 (to Tables 14 A–C)

Symmetrical trapezoidal mid-span line load

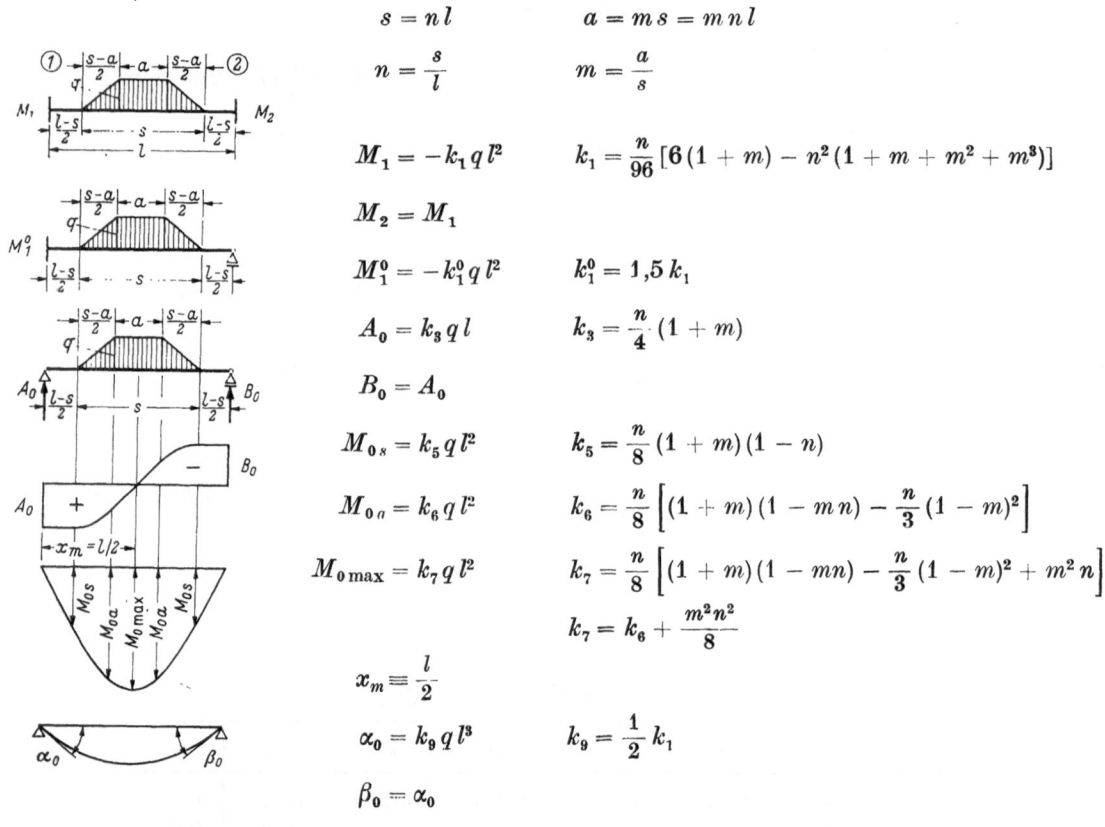

$$s = nl \qquad a = ms = mnl$$

$$n = \frac{s}{l} \qquad m = \frac{a}{s}$$

$$M_1 = -k_1 q l^2 \qquad k_1 = \frac{n}{96}\left[6(1+m) - n^2(1+m+m^2+m^3)\right]$$

$$M_2 = M_1$$

$$M_1^0 = -k_1^0 q l^2 \qquad k_1^0 = 1{,}5\,k_1$$

$$A_0 = k_3 q l \qquad k_3 = \frac{n}{4}(1+m)$$

$$B_0 = A_0$$

$$M_{0s} = k_5 q l^2 \qquad k_5 = \frac{n}{8}(1+m)(1-n)$$

$$M_{0a} = k_6 q l^2 \qquad k_6 = \frac{n}{8}\left[(1+m)(1-mn) - \frac{n}{3}(1-m)^2\right]$$

$$M_{0\max} = k_7 q l^2 \qquad k_7 = \frac{n}{8}\left[(1+m)(1-mn) - \frac{n}{3}(1-m)^2 + m^2 n\right]$$

$$k_7 = k_6 + \frac{m^2 n^2}{8}$$

$$x_m \equiv \frac{l}{2}$$

$$\alpha_0 = k_9 q l^3 \qquad k_9 = \frac{1}{2}\,k_1$$

$$\beta_0 = \alpha_0$$

Table 14A

n		m 0	0.1	0.2	0.3	0.4	0.5	0.6	0.7	0.8	0.9	1.0	n
0	k_1	0	0	0	0	0	0	0	0	0	0	0	k_1 0
	k_1^0	0	0	0	0	0	0	0	0	0	0	0	k_1^0
0.05	k_1	0.003	0.003	0.004	0.004	0.004	0.005	0.005	0.005	0.006	0.006	0.006	k_1 0.05
	k_1^0	.005	.005	.006	.006	.007	.007	.007	.008	.008	.009	.009	k_1^0
0.1	k_1	.006	.007	.007	.008	.009	.009	.010	.011	.011	.012	.012	k_1 0.1
	k_1^0	.009	.010	.011	.012	.013	.014	.015	.016	.017	.018	.019	k_1^0
0.15	k_1	.009	.010	.011	.012	.013	.014	.015	.016	.017	.018	.019	k_1 0.15
	k_1^0	.014	.015	.017	.018	.020	.021	.022	.024	.025	.027	.028	k_1^0
0.2	k_1	.012	.014	.015	.016	.017	.019	.020	.021	.022	.023	.025	k_1 0.2
	k_1^0	.019	.020	.022	.024	.026	.028	.030	.032	.033	.035	.037	k_1^0
0.25	k_1	.015	.017	.019	.020	.022	.023	.025	.026	.028	.029	.031	k_1 0.25
	k_1^0	.023	.026	.028	.030	.032	.035	.037	.039	.041	.044	.046	k_1^0
0.3	k_1	.019	.020	.022	.024	.026	.028	.029	.031	.033	.035	.036	k_1 0.3
	k_1^0	.028	.030	.033	.036	.039	.041	.044	.047	.049	.052	.055	k_1^0
0.35	k_1	.021	.024	.026	.028	.030	.032	.034	.036	.038	.040	.042	k_1 0.35
	k_1^0	.032	.035	.039	.042	.045	.048	.051	.054	.057	.060	.063	k_1^0
0.4	k_1	.024	.027	.029	.032	.034	.036	.039	.041	.043	.045	.047	k_1 0.4
	k_1^0	.037	.040	.044	.047	.051	.054	.058	.061	.065	.068	.071	k_1^0
0.45	k_1	.027	.030	.033	.035	.038	.040	.043	.045	.048	.050	.052	k_1 0.45
	k_1^0	.041	.045	.049	.053	.057	.061	.064	.068	.072	.075	.079	k_1^0
0.5	k_1	.030	.033	.036	.039	.042	.044	.047	.050	.052	.055	.057	k_1 0.5
	k_1^0	.045	.049	.054	.058	.062	.067	.071	.075	.079	.082	.086	k_1^0
0.55	k_1	.033	.036	.039	.042	.045	.048	.051	.054	.057	.059	.062	k_1 0.55
	k_1^0	.049	.054	.059	.063	.048	.072	.077	.081	.085	.089	.093	k_1^0
0.6	k_1	.035	.039	.042	.046	.049	.052	.055	.058	.061	.064	.066	k_1 0.6
	k_1^0	.053	.058	.063	.068	.073	.078	.083	.087	.091	.095	.099	k_1^0
0.65	k_1	.038	.042	.045	.049	.052	.056	.059	.062	.065	.067	.070	k_1 0.65
	k_1^0	.057	.062	.068	.073	.078	.083	.088	.093	.097	.101	.105	k_1^0
0.7	k_1	.040	.044	.048	.052	.055	.059	.062	.065	.068	.071	.073	k_1 0.7
	k_1^0	.060	.066	.072	.078	.083	.088	.093	.098	.102	.106	.110	k_1^0
0.75	k_1	.042	.047	.051	.055	.058	.062	.065	.069	.071	.074	.076	k_1 0.75
	k_1^0	.064	.070	.076	.082	.088	.093	.098	.103	.107	.111	.114	k_1^0
0.8	k_1	.045	.049	.053	.057	.061	.065	.068	.071	.074	.077	.079	k_1 0.8
	k_1^0	.067	.074	.080	.086	.092	.098	.103	.107	.111	.115	.118	k_1^0
0.85	k_1	.047	.051	.056	.060	.064	.068	.071	.074	.077	.079	.081	k_1 0.85
	k_1^0	.071	.077	.084	.090	.096	.102	.107	.111	.115	.118	.121	k_1^0
0.9	k_1	.049	.053	.058	.062	.066	.070	.073	.076	.079	.081	.082	k_1 0.9
	k_1^0	.073	.080	.087	.094	.100	.105	.110	.115	.118	.121	.123	k_1^0
0.95	k_1	.050	.055	.060	.065	.069	.072	.076	.078	.081	.082	.083	k_1 0.95
	k_1^0	.076	.083	.091	.097	.103	.108	.113	.117	.121	.123	.125	k_1^0
1.0	k_1	.052	.057	.062	.066	.071	.074	.077	.080	.082	.083	.083	k_1 1.0
	k_1^0	0.078	0.086	0.093	0.100	0.106	0.111	0.116	0.120	0.123	0.124	0.125	k_1^0

Table 14 B   – 44 –

Table 14B

n		m 0	0.1	0.2	0.3	0.4	0.5	0.6	0.7	0.8	0.9	1.0		n
0	k_3	0	0	0	0	0	0	0	0	0	0	0	k_3	0
	k_9												k_9	
0.05	k_3	0.013	0.014	0.015	0.016	0.018	0.019	0.020	0.021	0.023	0.024	0.025	k_3	0.05
	k_9	.002	.002	.002	.002	.002	.002	.003	.003	.003	.003	.003	k_9	
0.1	k_3	.025	.028	.030	.033	.035	.038	.040	.043	.045	.048	.050	k_3	0.1
	k_9	.003	.003	.004	.004	.004	.005	.005	.005	.006	.006	.006	k_9	
0.15	k_3	.038	.041	.045	.049	.053	.056	.060	.064	.068	.071	.075	k_3	0.15
	k_9	.005	.005	.006	.006	.007	.007	.007	.008	.008	.009	.009	k_9	
0.2	k_3	.050	.055	.060	.065	.070	.075	.080	.085	.090	.095	.100	k_3	0.2
	k_9	.006	.007	.007	.008	.009	.009	.010	.011	.011	.012	.012	k_9	
0.25	k_3	.063	.069	.075	.081	.088	.094	.100	.106	.113	.119	.125	k_3	0.25
	k_9	.008	.009	.009	.010	.011	.012	.012	.013	.014	.015	.015	k_9	
0.3	k_3	.075	.083	.090	.098	.105	.113	.120	.128	.135	.143	.150	k_3	0.3
	k_9	.009	.010	.011	.012	.013	.014	.015	.016	.016	.017	.018	k_9	
0.35	k_3	.088	.096	.105	.114	.123	.131	.140	.149	.158	.166	.175	k_3	0.35
	k_9	.011	.012	.013	.014	.015	.016	.017	.018	.019	.020	.021	k_9	
0.4	k_3	.100	.110	.120	.130	.140	.150	.160	.170	.180	.190	.200	k_3	0.4
	k_9	.012	.013	.015	.016	.017	.018	.019	.020	.022	.023	.024	k_9	
0.45	k_3	.113	.124	.135	.146	.158	.169	.180	.191	.203	.214	.225	k_3	0.45
	k_9	.014	.015	.016	.018	.019	.020	.021	.023	.024	.025	.026	k_9	
0.5	k_3	.125	.138	.150	.163	.175	.188	.200	.213	.225	.238	.250	k_3	0.5
	k_9	.015	.016	.018	.019	.021	.022	.024	.025	.026	.027	.029	k_9	
0.55	k_3	.138	.151	.165	.179	.193	.206	.220	.234	.248	.261	.275	k_3	0.55
	k_9	.016	.018	.020	.021	.023	.024	.026	.027	.028	.030	.031	k_9	
0.6	k_3	.150	.165	.180	.195	.210	.225	.240	.255	.270	.285	.300	k_3	0.6
	k_9	.018	.019	.021	.023	.024	.026	.028	.029	.030	.032	.033	k_9	
0.65	k_3	.163	.179	.195	.211	.228	.244	.260	.276	.293	.309	.325	k_3	0.65
	k_9	.019	.021	.023	.024	.026	.028	.029	.031	.032	.034	.035	k_9	
0.7	k_3	.175	.193	.210	.228	.245	.263	.280	.298	.315	.333	.350	k_3	0.7
	k_9	.020	.022	.024	.026	.028	.029	.031	.033	.034	.035	.037	k_9	
0.75	k_3	.188	.206	.225	.244	.263	.281	.300	.319	.338	.356	.375	k_3	0.75
	k_9	.021	.023	.025	.027	.029	.031	.033	.034	.036	.037	.038	k_9	
0.8	k_3	.200	.220	.240	.260	.280	.300	.320	.340	.360	.380	.400	k_3	0.8
	k_9	.022	.025	.027	.029	.031	.033	.034	.036	.037	.038	.039	k_9	
0.85	k_3	.213	.234	.255	.276	.298	.319	.340	.361	.383	.404	.425	k_3	0.85
	k_9	.023	.026	.028	.030	.032	.034	.036	.037	.038	.039	.040	k_9	
0.9	k_3	.225	.248	.270	.293	.315	.338	.360	.383	.405	.428	.450	k_3	0.9
	k_9	.024	.027	.029	.031	.033	.035	.037	.038	.039	.040	.041	k_9	
0.95	k_3	.238	.261	.285	.309	.333	.356	.380	.404	.428	.451	.475	k_3	0.95
	k_9	.025	.028	.030	.032	.034	.036	.038	.039	.040	.041	.042	k_9	
1.0	k_3	.250	.275	.300	.325	.350	.375	.400	.425	.450	.475	.500	k_3	1.0
	k_9	0.026	0.029	0.031	0.033	0.035	0.037	0.039	0.040	0.041	0.041	0.042	k_9	

Table 14C

n		m									n
		0.1	0.2	0.3	0.4	0.5	0.6	0.7	0.8	0.9	
0	k_5	0	0	0	0	0	0	0	0	0	k_5
	k_6										k_6 0
	k_7										k_7
0.05	k_5	0.007	0.007	0.008	0.008	0.009	0.010	0.010	0.011	0.011	k_5
	k_6	.007	.007	.008	.009	.009	.010	.010	.011	.011	k_6 0.05
	k_7	.007	.007	.008	.009	.009	.010	.010	.011	.012	k_7
0.1	k_5	.012	.014	.015	.016	.017	.018	.019	.020	.021	k_5
	k_6	.013	.014	.015	.017	.018	.019	.020	.021	.022	k_6 0.1
	k_7	.013	.014	.016	.017	.018	.019	.020	.021	.023	k_7
0.15	k_5	.018	.019	.021	.022	.024	.026	.027	.029	.030	k_5
	k_6	.019	.021	.023	.024	.025	.027	.028	.030	.031	k_6 0.15
	k_7	.019	.021	.023	.024	.026	.028	.030	.031	.033	k_7
0.2	k_5	.022	.024	.026	.028	.030	.032	.034	.036	.038	k_5
	k_6	.025	.027	.029	.031	.033	.035	.036	.038	.039	k_6 0.2
	k_7	.025	.027	.030	.032	.034	.036	.039	.041	.043	k_7
0.25	k_5	.026	.028	.030	.033	.035	.038	.040	.042	.045	k_5
	k_6	.031	.033	.035	.038	.040	.041	.043	.045	.046	k_6 0.25
	k_7	.031	.034	.036	.039	0.42	.044	.047	.050	.052	k_7
0.3	k_5	.029	.032	.034	.037	.039	.042	.045	.047	.050	k_5
	k_6	.036	.039	.041	.044	.046	.048	.049	.051	.052	k_6 0.3
	k_7	.036	.039	.042	.045	.049	.052	.055	.058	.061	k_7
0.35	k_5	.031	.034	.037	.040	.043	.046	.048	.051	.054	k_5
	k_6	.041	.044	.047	.049	.051	.053	.055	.056	.057	k_6 0.35
	k_7	.042	.045	.048	.052	.055	.059	.062	.066	.069	k_7
0.4	k_5	.033	.036	.039	.042	.045	.048	.051	.054	.057	k_5
	k_6	.046	.049	.052	.054	.056	.058	.059	.060	.061	k_6 0.4
	k_7	.046	.050	.053	.057	.061	.065	.069	.073	.077	k_7
0.45	k_5	.034	.037	.040	.043	.046	.050	.053	.056	.059	k_5
	k_6	.051	.054	.056	.059	.061	.062	.063	.064	.063	k_6 0.45
	k_7	.051	.055	.059	.063	.067	.071	.076	.080	.084	k_7
0.5	k_5	.034	.038	.041	.044	.047	.050	.053	.056	.059	k_5
	k_6	.055	.058	.060	.063	.064	.066	.066	.066	.065	k_6 0.5
	k_7	.055	.059	.063	.068	.072	.077	.082	.086	.090	k_7
0.55	k_5	.034	.037	.040	.043	.046	.050	.053	.056	.059	k_5
	k_6	.059	.062	.064	.066	.068	.069	.069	.068	.066	k_6 0.55
	k_7	.060	.063	.068	.072	.077	.082	.087	.092	.096	k_7
0.6	k_5	.033	.036	.039	.042	.045	.048	.051	.054	.057	k_5
	k_6	.063	.065	.068	.069	.070	.071	.070	.068	.065	k_6 0.6
	k_7	.063	.067	.072	.076	.082	.087	.092	.097	.101	k_7
0.65	k_5	.031	.034	.037	.040	.043	.046	.048	.051	.054	k_5
	k_6	.066	.069	.070	.072	.072	.072	.071	.068	.063	k_6 0.65
	k_7	.067	.071	.075	.080	.086	.091	.097	.102	.106	k_7
0.7	k_5	.029	.032	.034	.037	.039	.042	.045	.047	.050	k_5
	k_6	.070	.071	.073	.074	.074	.073	.071	.067	.061	k_6 0.7
	k_7	.070	.074	.078	.084	.089	.095	.101	.106	.110	k_7
0.75	k_5	.026	.028	.030	.033	.035	.038	.040	.042	.045	k_5
	k_6	.072	.074	.075	.075	.075	.073	.070	.064	.057	k_6 0.75
	k_7	.073	.077	.081	.087	.092	.098	.104	.109	.114	k_7

4*

To be continued Table 14C on p. 46

Table 14 C — 46 —

Table 14C (Continuation)

n		m									n	
		0.1	0.2	0.3	0.4	0.5	0.6	0.7	0.8	0.9		
	k_5	0.022	0.024	0.026	0.028	0.030	0.032	0.034	0.036	0.038	k_5	
0.8	k_6	.075	.076	.077	.076	.075	.072	.068	.061	.052	k_6	0.8
	k_7	.076	.079	.084	.089	.095	.101	.107	.113	.117	k_7	
	k_5	.018	.019	.021	.022	.024	.026	.027	.029	.030	k_5	
0.85	k_6	.077	.078	.078	.077	.075	.071	.065	.057	.046	k_6	0.85
	k_7	.078	.082	.086	.091	.097	.103	.110	.115	.120	k_7	
	k_5	.012	.014	.015	.016	.017	.018	.019	.020	.021	k_5	
0.9	k_6	.080	.080	.079	.077	.074	.069	.062	.052	.039	k_6	0.9
	k_7	.081	.084	.088	.093	.099	.105	.112	.117	.121	k_7	
	k_5	.007	.007	.008	.008	.009	.010	.010	.011	.011	k_5	
0.95	k_6	.081	.081	.079	.077	.072	.066	.058	.046	.031	k_6	0.95
	k_7	.083	.085	.089	.095	.101	.107	.113	.119	.123	k_7	
	k_5	0	0	0	0	0	0	0	0	0	k_5	
1.0	k_6	.083	.082	.079	.076	.070	.063	.053	.040	.022	k_6	1.0
	k_7	0.084	0.087	0.091	0.096	.0102	0.108	0.114	0.120	0.124	k_7	

Loading 15 (to Table 15)

Bound trapezoidal load

$$s = n\,l$$

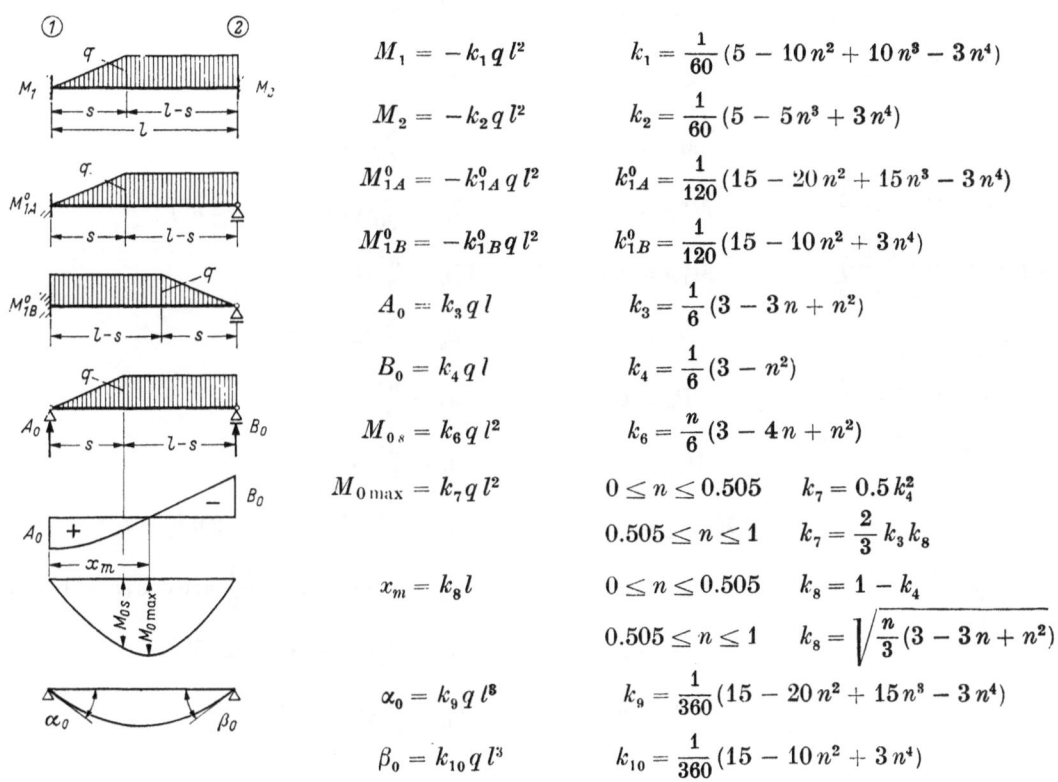

$$M_1 = -k_1 q\,l^2 \qquad k_1 = \frac{1}{60}(5 - 10\,n^2 + 10\,n^3 - 3\,n^4)$$

$$M_2 = -k_2 q\,l^2 \qquad k_2 = \frac{1}{60}(5 - 5\,n^3 + 3\,n^4)$$

$$M_{1A}^0 = -k_{1A}^0 q\,l^2 \qquad k_{1A}^0 = \frac{1}{120}(15 - 20\,n^2 + 15\,n^3 - 3\,n^4)$$

$$M_{1B}^0 = -k_{1B}^0 q\,l^2 \qquad k_{1B}^0 = \frac{1}{120}(15 - 10\,n^2 + 3\,n^4)$$

$$A_0 = k_3 q\,l \qquad k_3 = \frac{1}{6}(3 - 3\,n + n^2)$$

$$B_0 = k_4 q\,l \qquad k_4 = \frac{1}{6}(3 - n^2)$$

$$M_{0s} = k_6 q\,l^2 \qquad k_6 = \frac{n}{6}(3 - 4\,n + n^2)$$

$$M_{0\max} = k_7 q\,l^2 \qquad 0 \le n \le 0.505 \quad k_7 = 0.5\,k_4^2$$

$$0.505 \le n \le 1 \quad k_7 = \frac{2}{3}\,k_3 k_8$$

$$x_m = k_8 l \qquad 0 \le n \le 0.505 \quad k_8 = 1 - k_4$$

$$0.505 \le n \le 1 \quad k_8 = \sqrt{\frac{n}{3}(3 - 3\,n + n^2)}$$

$$\alpha_0 = k_9 q\,l^3 \qquad k_9 = \frac{1}{360}(15 - 20\,n^2 + 15\,n^3 - 3\,n^4)$$

$$\beta_0 = k_{10} q\,l^3 \qquad k_{10} = \frac{1}{360}(15 - 10\,n^2 + 3\,n^4)$$

Table 15

n	k_1	k_2	k_{1A}^0	k_{1B}^0	k_3	k_4	k_6	k_7	k_8	k_9	k_{10}	n
0	0.083	0.083	0.125	0.125	0.500	0.500	0.000	0.125	0.500	0.042	0.042	0
0.05	.083	.083	.125	.125	.475	.500	.023	.125	.500	.042	.042	0.05
0.1	.082	.083	.123	.124	.452	.498	044	.124	.502	.041	.041	0.1
0.15	.080	.083	.122	.123	.429	.496	.061	.123	.504	.041	.041	0.15
0.2	.078	.083	.119	.122	.407	.493	.075	.122	.507	.040	.041	0.2
0.25	.075	.082	.116	.120	.385	.490	.086	.120	.510	.039	.040	0.25
0.3	.072	.081	.113	.118	.365	.485	.095	.118	.515	.038	.039	0.3
0.35	.069	.081	.110	.115	.345	.480	.100	.115	.520	.037	.038	0.35
0.4	.066	.079	.106	.112	.327	.473	.104	.112	.527	.035	.037	0.4
0.45	.063	.078	.102	.109	.309	.466	.105	.109	.534	.034	.036	0.45
0.5	.059	.076	.097	.106	.292	.458	.104	.105	.542	.032	.035	0.5
0.55	.056	.074	.093	.102	.275	.450	.101	.101	.550	.031	.034	0.55
0.6	.053	.072	.089	.098	.260	.440	.096	.097	.559	.030	.033	0.6
0.65	.050	.069	.084	.094	.245	.430	.089	.092	.565	.028	.031	0.65
0.7	.047	.067	.080	.090	.232	.418	.081	.088	.570	.027	.030	0.7
0.75	.044	.064	.076	.086	.219	.406	.070	.084	.573	.025	.029	0.75
0.8	.042	.061	.072	.082	.207	.393	.059	.079	.575	.024	.027	0.8
0.85	.039	.058	.068	.078	.195	.380	.046	.075	.576	.023	.026	0.85
0.9	.037	.055	.065	.074	.185	.365	.032	.071	.577	.022	.025	0.9
0.95	.035	.053	.061	.070	.175	.350	.016	.068	.577	.020	.023	0.95
1.0	.033	.050	.058	.067	.167	.333	0.000	.064	.577	.019	.022	1.0

Table 16   – 48 –

Loading 16 (to Table 16)

Bound load according to a quadratic parabola

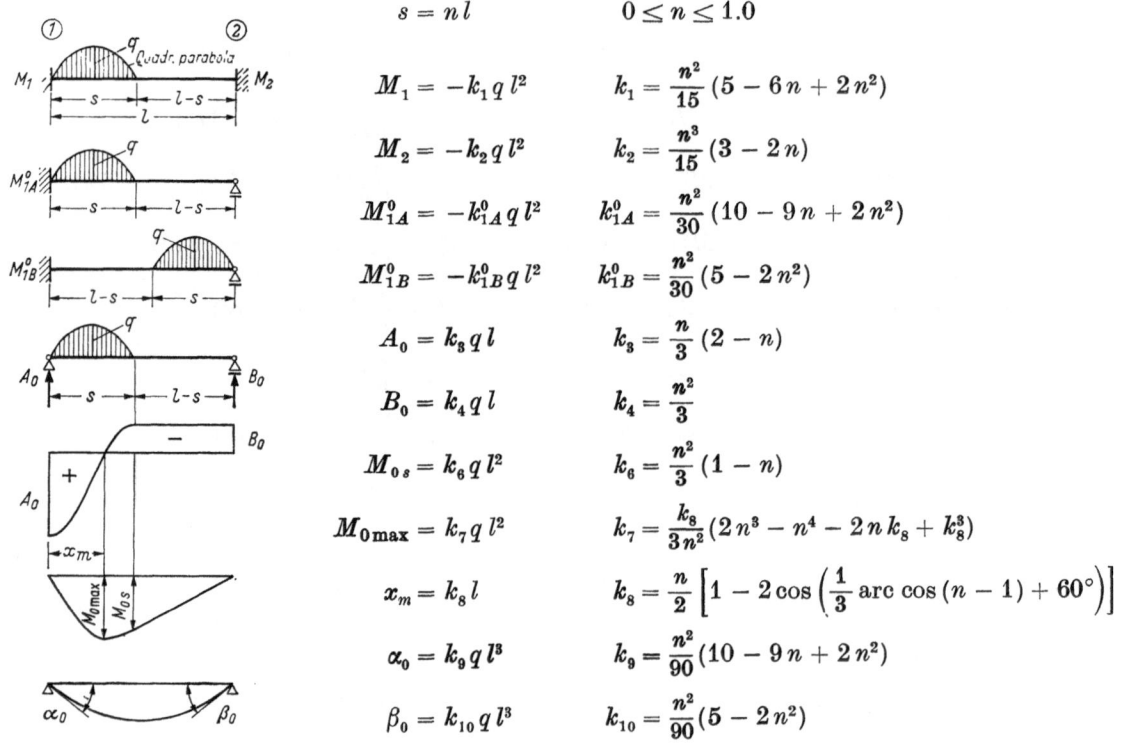

$$s = n\,l \qquad\qquad 0 \leq n \leq 1.0$$

$$M_1 = -k_1\,q\,l^2 \qquad k_1 = \frac{n^2}{15}(5 - 6\,n + 2\,n^2)$$

$$M_2 = -k_2\,q\,l^2 \qquad k_2 = \frac{n^3}{15}(3 - 2\,n)$$

$$M_{1A}^0 = -k_{1A}^0\,q\,l^2 \qquad k_{1A}^0 = \frac{n^2}{30}(10 - 9\,n + 2\,n^2)$$

$$M_{1B}^0 = -k_{1B}^0\,q\,l^2 \qquad k_{1B}^0 = \frac{n^2}{30}(5 - 2\,n^2)$$

$$A_0 = k_3\,q\,l \qquad k_3 = \frac{n}{3}(2 - n)$$

$$B_0 = k_4\,q\,l \qquad k_4 = \frac{n^2}{3}$$

$$M_{0s} = k_6\,q\,l^2 \qquad k_6 = \frac{n^2}{3}(1 - n)$$

$$M_{0\max} = k_7\,q\,l^2 \qquad k_7 = \frac{k_8}{3n^2}(2\,n^3 - n^4 - 2\,n\,k_8 + k_8^3)$$

$$x_m = k_8\,l \qquad k_8 = \frac{n}{2}\left[1 - 2\cos\left(\frac{1}{3}\arccos(n-1) + 60°\right)\right]$$

$$\alpha_0 = k_9\,q\,l^3 \qquad k_9 = \frac{n^2}{90}(10 - 9\,n + 2\,n^2)$$

$$\beta_0 = k_{10}\,q\,l^3 \qquad k_{10} = \frac{n^2}{90}(5 - 2\,n^2)$$

Table 16

n	k_1	k_2	k_{1A}^0	k_{1B}^0	k_3	k_4	k_6	k_7	k_8	k_9	k_{10}	n
0	0	0	0	0	0	0	0	0	0	0	0	0
0.05	0.001	0.000	0.001	0.000	0.033	0.001	0.001	0.001	0.045	0.000	0.000	0.05
0.1	.003	.000	.003	.002	.063	.003	.003	.003	.086	.001	.001	0.1
0.15	.006	.001	.007	.004	.093	.008	.006	.006	.125	.002	.001	0.15
0.2	.010	.001	.011	.007	.120	.013	.011	.011	.161	.004	.002	0.2
0.25	.015	.003	.016	.010	.146	.021	.016	.016	.195	.005	.003	0.25
0.3	.020	.004	.022	.014	.170	.030	.021	.022	.227	.007	.005	0.3
0.35	.026	.007	.029	.019	.193	.041	.027	.029	.257	.010	.006	0.35
0.4	.031	.009	.036	.025	.213	.053	.032	.036	.285	.012	.008	0.4
0.45	.037	.013	.043	.031	.233	.068	.037	.043	.312	.014	.010	0.45
0.5	.042	.017	.050	.038	.250	.083	.042	.050	.337	.017	.013	0.5
0.55	.046	.021	.057	.044	.266	.101	.045	.058	.360	.019	.015	0.55
0.6	.051	.026	.064	.051	.280	.120	.048	.065	.382	.021	.017	0.6
0.65	.055	.031	.070	.059	.293	.141	.049	.072	.402	.023	.020	0.65
0.7	.058	.037	.076	.066	.303	.163	.049	.078	.421	.025	.022	0.7
0.75	.061	.042	.082	.073	.313	.188	.047	.084	.438	.027	.024	0.75
0.8	.063	,048	.087	.079	.320	.213	.043	.086	.454	.029	.026	0.8
0.85	.065	.053	.091	.086	.326	.241	.036	.094	.468	.030	.029	0.85
0.9	.066	.058	.095	.091	.330	.270	.027	.098	.480	.032	.030	0.9
0.95	.066	.063	.098	.096	.333	.301	.015	.102	.491	.033	.032	0.95
1.0	0.067	0.067	0.100	0.100	0.333	0.333	0.000	0.104	0.500	0.033	0.033	1.0

Loading 17 (to Table 17)

Symmetrical load according to a quadratic parabola

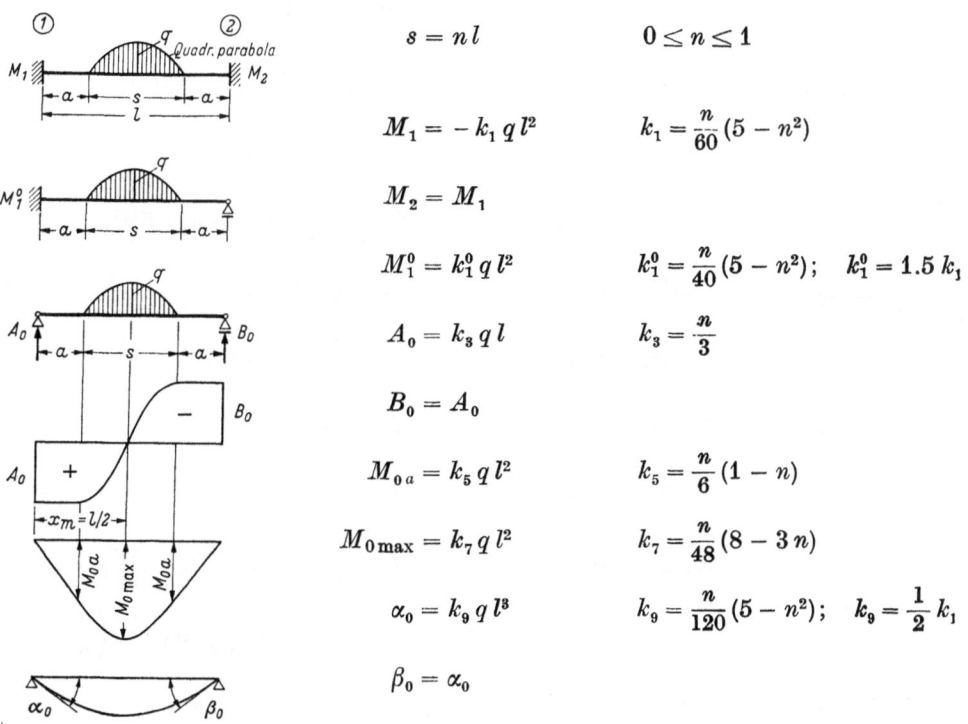

$$s = n\,l \qquad\qquad 0 \le n \le 1$$

$$M_1 = -\,k_1\,q\,l^2 \qquad k_1 = \frac{n}{60}\,(5 - n^2)$$

$$M_2 = M_1$$

$$M_1^0 = k_1^0\,q\,l^2 \qquad k_1^0 = \frac{n}{40}\,(5 - n^2);\quad k_1^0 = 1.5\,k_1$$

$$A_0 = k_3\,q\,l \qquad k_3 = \frac{n}{3}$$

$$B_0 = A_0$$

$$M_{0\,a} = k_5\,q\,l^2 \qquad k_5 = \frac{n}{6}\,(1 - n)$$

$$M_{0\,\max} = k_7\,q\,l^2 \qquad k_7 = \frac{n}{48}\,(8 - 3\,n)$$

$$\alpha_0 = k_9\,q\,l^3 \qquad k_9 = \frac{n}{120}\,(5 - n^2);\quad k_9 = \frac{1}{2}\,k_1$$

$$\beta_0 = \alpha_0$$

Table 17

n	k_1	k_1^0	k_3	k_5	k_7	k_9	n
0	0	0	0	0	0	0	0
0.05	0.004	0.006	0.017	0.008	0.008	0.002	0.05
0.1	.008	.012	.033	.015	.016	.004	0.1
0.15	.012	.019	.050	.021	.024	.006	0.15
0.2	.017	.025	.067	.027	.031	.008	0.2
0.25	.021	.031	.083	.031	.038	.010	0.25
0.3	.025	.037	.100	.035	.044	.012	0.3
0.35	.028	.043	.117	.038	.051	.014	0.35
0.4	.032	.048	.133	.040	.057	.016	0.4
0.45	.036	.054	.150	.041	.062	.018	0.45
0.5	.040	.059	.167	.042	.068	.020	0.5
0.55	.043	.065	.183	.041	.073	.022	0.55
0.6	.046	.070	.200	.040	.078	.023	0.6
0.65	.050	.074	.217	.038	.082	.025	0.65
0.7	.053	.079	.233	.035	.086	.026	0.7
0.75	.055	.083	.250	.031	.090	.028	0.75
0.8	.058	.087	.267	.027	.093	.029	0.8
0.85	.061	.091	.283	.021	.097	.030	0.85
0.9	.063	.094	.300	.015	.099	.031	0.9
0.95	.065	.097	.317	.008	.102	.032	0.95
1.0	0.067	0.100	0.333	0.000	0.104	0.033	1.0

Loading 18 (to Table 18)

Bound load according to a quadratic parabola

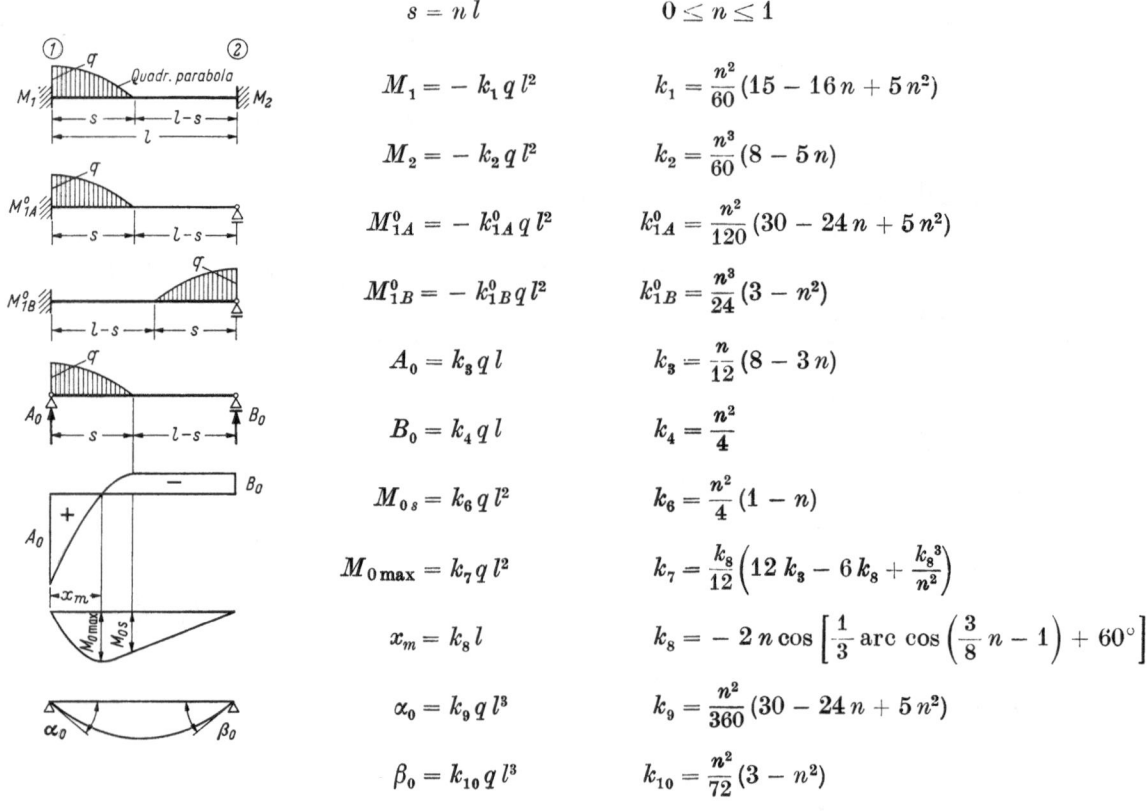

$$s = n\,l \qquad\qquad 0 \le n \le 1$$

$$M_1 = -\,k_1\,q\,l^2 \qquad\qquad k_1 = \frac{n^2}{60}\,(15 - 16\,n + 5\,n^2)$$

$$M_2 = -\,k_2\,q\,l^2 \qquad\qquad k_2 = \frac{n^3}{60}\,(8 - 5\,n)$$

$$M_{1A}^0 = -\,k_{1A}^0\,q\,l^2 \qquad\qquad k_{1A}^0 = \frac{n^2}{120}\,(30 - 24\,n + 5\,n^2)$$

$$M_{1B}^0 = -\,k_{1B}^0\,q\,l^2 \qquad\qquad k_{1B}^0 = \frac{n^3}{24}\,(3 - n^2)$$

$$A_0 = k_3\,q\,l \qquad\qquad k_3 = \frac{n}{12}\,(8 - 3\,n)$$

$$B_0 = k_4\,q\,l \qquad\qquad k_4 = \frac{n^2}{4}$$

$$M_{0s} = k_6\,q\,l^2 \qquad\qquad k_6 = \frac{n^2}{4}\,(1 - n)$$

$$M_{0\,\text{max}} = k_7\,q\,l^2 \qquad\qquad k_7 = \frac{k_8}{12}\left(12\,k_3 - 6\,k_8 + \frac{k_8^3}{n^2}\right)$$

$$x_m = k_8\,l \qquad\qquad k_8 = -\,2\,n \cos\left[\frac{1}{3}\,\text{arc}\cos\left(\frac{3}{8}\,n - 1\right) + 60°\right]$$

$$\alpha_0 = k_9\,q\,l^3 \qquad\qquad k_9 = \frac{n^2}{360}\,(30 - 24\,n + 5\,n^2)$$

$$\beta_0 = k_{10}\,q\,l^3 \qquad\qquad k_{10} = \frac{n^2}{72}\,(3 - n^2)$$

Table 18

n	k_1	k_2	k_{1A}^0	k_{1B}^0	k_3	k_4	k_6	k_7	k_8	k_9	k_{10}	n
0	0	0	0	0	0	0	0	0	0	0	0	0
0.05	0.001	0.000	0.001	0.000	0.033	0.001	0.001	0.001	0.044	0.000	0.000	0.05
0.1	.002	.000	.002	.001	.064	.003	.002	.002	.084	.001	.000	0.1
0.15	.005	.000	.005	.003	.094	.006	.005	.005	.120	.002	.001	0.15
0.2	.008	.001	.008	.005	.123	.010	.008	.008	.153	.003	.002	0.2
0.25	.012	.002	.013	.008	.151	.016	.012	.012	.185	.004	.003	0.25
0.3	.016	.003	.017	.011	.178	.023	.016	.017	.214	.006	.004	0.3
0.35	.020	.004	.023	.015	.203	.031	.020	.022	.241	.008	.005	0.35
0.4	.025	.006	.028	.019	.227	.040	.024	.028	.266	.009	.006	0.4
0.45	.030	.009	.034	.024	.249	.051	.028	.033	.289	.011	.008	0.45
0.5	.034	.011	.040	.029	.271	.063	.031	.039	.311	.013	.010	0.5
0.55	.039	.015	.046	.034	.291	.076	.034	.045	.331	.015	.011	0.55
0.6	.043	.018	.052	.040	.310	.090	.036	.051	.350	.017	.013	0.6
0.65	.047	.022	.058	.045	.328	.106	.037	.057	.367	.019	.015	0.65
0.7	.051	.026	.064	.051	.344	.123	.037	.062	.382	.021	.017	0.7
0.75	.054	.030	.069	.057	.359	.141	.035	.068	.396	.023	.019	0.75
0.8	.058	.034	.075	.063	.373	.160	.032	.073	.409	.025	.021	0.8
0.85	.061	.038	.080	.069	.386	.181	.027	.078	.420	.027	.023	0.85
0.9	.063	.043	.084	.074	.398	.203	.020	.082	.430	.028	.025	0.9
0.95	.065	.046	.088	.079	.408	.226	.011	.086	.439	.029	.026	0.95
1.0	0.067	0.050	0.092	0.083	0.417	0.250	0.000	0.090	0.446	0.031	0.028	1.0

Table 19

Loading 19 (to Table 19)

Bound load according to a quadratic parabola

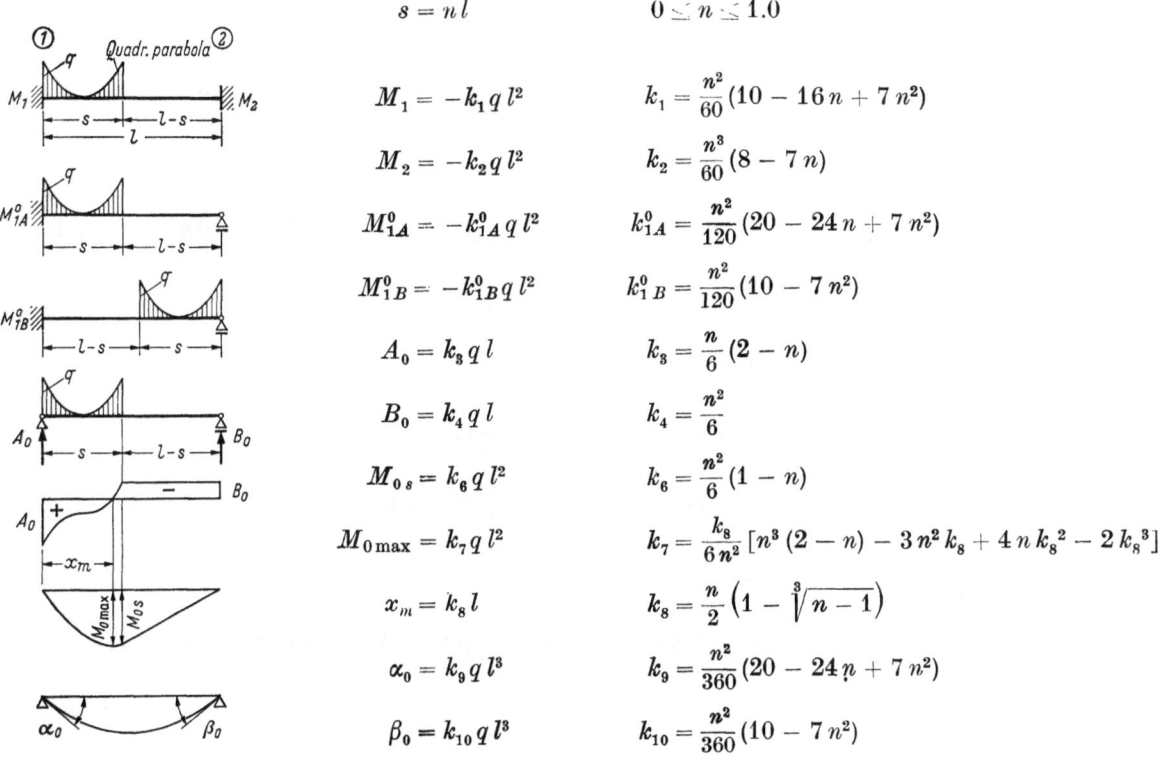

$$s = n\,l \qquad\qquad 0 \leqq n \leqq 1.0$$

$$M_1 = -k_1\,q\,l^2 \qquad\qquad k_1 = \frac{n^2}{60}(10 - 16\,n + 7\,n^2)$$

$$M_2 = -k_2\,q\,l^2 \qquad\qquad k_2 = \frac{n^3}{60}(8 - 7\,n)$$

$$M_{1A}^0 = -k_{1A}^0\,q\,l^2 \qquad\qquad k_{1A}^0 = \frac{n^2}{120}(20 - 24\,n + 7\,n^2)$$

$$M_{1B}^0 = -k_{1B}^0\,q\,l^2 \qquad\qquad k_{1B}^0 = \frac{n^2}{120}(10 - 7\,n^2)$$

$$A_0 = k_3\,q\,l \qquad\qquad k_3 = \frac{n}{6}(2 - n)$$

$$B_0 = k_4\,q\,l \qquad\qquad k_4 = \frac{n^2}{6}$$

$$M_{0s} = k_6\,q\,l^2 \qquad\qquad k_6 = \frac{n^2}{6}(1 - n)$$

$$M_{0\,max} = k_7\,q\,l^2 \qquad\qquad k_7 = \frac{k_8}{6\,n^2}[n^3(2 - n) - 3\,n^2\,k_8 + 4\,n\,k_8{}^2 - 2\,k_8{}^3]$$

$$x_m = k_8\,l \qquad\qquad k_8 = \frac{n}{2}\left(1 - \sqrt[3]{n - 1}\right)$$

$$\alpha_0 = k_9\,q\,l^3 \qquad\qquad k_9 = \frac{n^2}{360}(20 - 24\,n + 7\,n^2)$$

$$\beta_0 = k_{10}\,q\,l^3 \qquad\qquad k_{10} = \frac{n^2}{360}(10 - 7\,n^2)$$

Table 19

n	k_1	k_2	k_{1A}^0	k_{1B}^0	k_3	k_4	k_6	k_7	k_8	k_9	k_{10}	n
0	0	0	0	0	0	0	0	0	0	0	0	0
0.05	0.000	0.000	0.000	0.000	0.016	0.000	0.000	0.001	0.050	0.000	0.000	0.05
0.1	.001	.000	.000	.001	.032	.002	.001	.001	.098	.000	.000	0.1
0.15	.003	.000	.003	.002	.046	.004	.003	.003	.146	.001	.001	0.15
0.2	.005	.001	.005	.003	.060	.007	.005	.005	.193	.002	.001	0.2
0.25	.007	.002	.008	.005	.073	.010	.008	.008	.239	.003	.002	0.25
0.3	.009	.003	.010	.007	.085	.015	.011	.011	.283	.003	.002	0.3
0.35	.011	.004	.013	.009	.096	.020	.013	.014	.327	.004	.003	0.35
0.4	.013	.006	.015	.012	.107	.027	.016	.016	.369	.005	.004	0.4
0.45	.014	.007	.018	.014	.116	.034	.019	.019	.409	.006	.005	0.45
0.5	.016	.009	.020	.017	.125	.042	.021	.022	.448	.007	.006	0.5
0.55	.017	.012	.022	.020	.133	.050	.023	.024	.486	.007	.007	0.55
0.6	.018	.014	.024	.022	.140	.060	.024	.026	.521	.008	.007	0.6
0.65	.018	.016	.026	.025	.146	.070	.025	.028	.554	.009	.008	0.65
0.7	.018	.018	.027	.027	.152	.082	.025	.029	.584	.009	.009	0.7
0.75	.018	.019	.028	.028	.156	.094	.023	.029	.611	.009	.009	0.75
0.8	.018	.020	.028	.029	.160	.107	.021	.029	.634	.009	.010	0.8
0.85	.018	.021	,028	.030	.163	.120	.018	.028	.651	.009	.010	0.85
0.9	.017	.021	.027	.029	.165	.135	.014	.026	.659	.009	.010	0.9
0.95	.017	.019	.026	.028	.166	.150	.008	.024	.650	.009	.009	0.95
1.0	0.017	0.017	0.025	0.025	0.167	0.167	0.000	0.021	0.500	0.008	0.008	1.0

Table 20   – 52 –

Loading 20 (to Table 20)

Symmetrical load according to a quadratic parabola

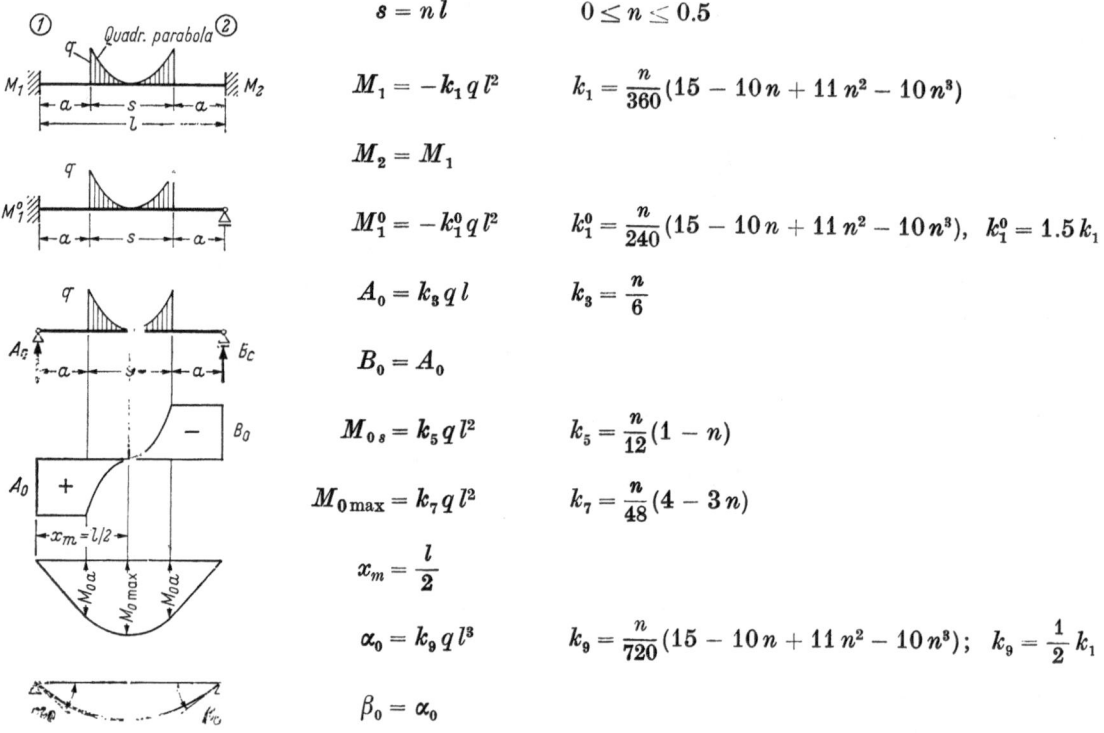

$$s = n\,l \qquad\qquad 0 \leq n \leq 0.5$$

$$M_1 = -k_1 q\, l^2 \qquad k_1 = \frac{n}{360}(15 - 10\,n + 11\,n^2 - 10\,n^3)$$

$$M_2 = M_1$$

$$M_1^0 = -k_1^0 q\, l^2 \qquad k_1^0 = \frac{n}{240}(15 - 10\,n + 11\,n^2 - 10\,n^3), \quad k_1^0 = 1.5\,k_1$$

$$A_0 = k_3 q\, l \qquad k_3 = \frac{n}{6}$$

$$B_0 = A_0$$

$$M_{0\,s} = k_5 q\, l^2 \qquad k_5 = \frac{n}{12}(1 - n)$$

$$M_{0\,max} = k_7 q\, l^2 \qquad k_7 = \frac{n}{48}(4 - 3\,n)$$

$$x_m = \frac{l}{2}$$

$$\alpha_0 = k_9 q\, l^3 \qquad k_9 = \frac{n}{720}(15 - 10\,n + 11\,n^2 - 10\,n^3); \quad k_9 = \frac{1}{2}\,k_1$$

$$\beta_0 = \alpha_0$$

Table 20

n	k_1	k_1^0	k_3	k_5	k_7	k_9	n
0	0	0	0	0	0	0	0
0.05	0.002	0.003	0.008	0.004	0.004	0.001	0.05
0.1	.004	.006	.017	.008	.008	.002	0.1
0.15	.006	.009	.025	.011	.011	.003	0.15
0.2	.007	.011	.033	.013	.014	.004	0.2
0.25	.009	.013	.042	.016	.017	.005	0.25
0.3	.011	.016	.050	.018	.019	.005	0.3
0.35	.012	.018	.058	.019	.022	.006	0.35
0.4	.013	.020	.067	.020	.023	.007	0.4
0.45	.015	.022	.075	.021	.025	.007	0.45
0.5	.016	.024	.083	.021	.026	.008	0.5
0.55	.017	.026	.092	.021	.027	.009	0.55
0.6	.018	.027	.100	.020	.028	.009	0.6
0.65	.019	.028	.108	.019	.028	.009	0.65
0.7	.019	.029	.117	.018	.028	.010	0.7
0.75	.020	.030	.125	.016	.027	.010	0.75
0.8	.020	.030	.133	.013	.027	.010	0.8
0.85	.020	.029	.142	.011	.026	.010	0.85
0.9	.019	.029	.150	.008	.024	.010	0.9
0.95	.018	.027	.158	.004	.023	.009	0.95
1.0	0.017	0.025	0.167	0.000	0.021	0.008	1.0

Loading 21 (to Table 21)

Bound load according to a quadratic parabola

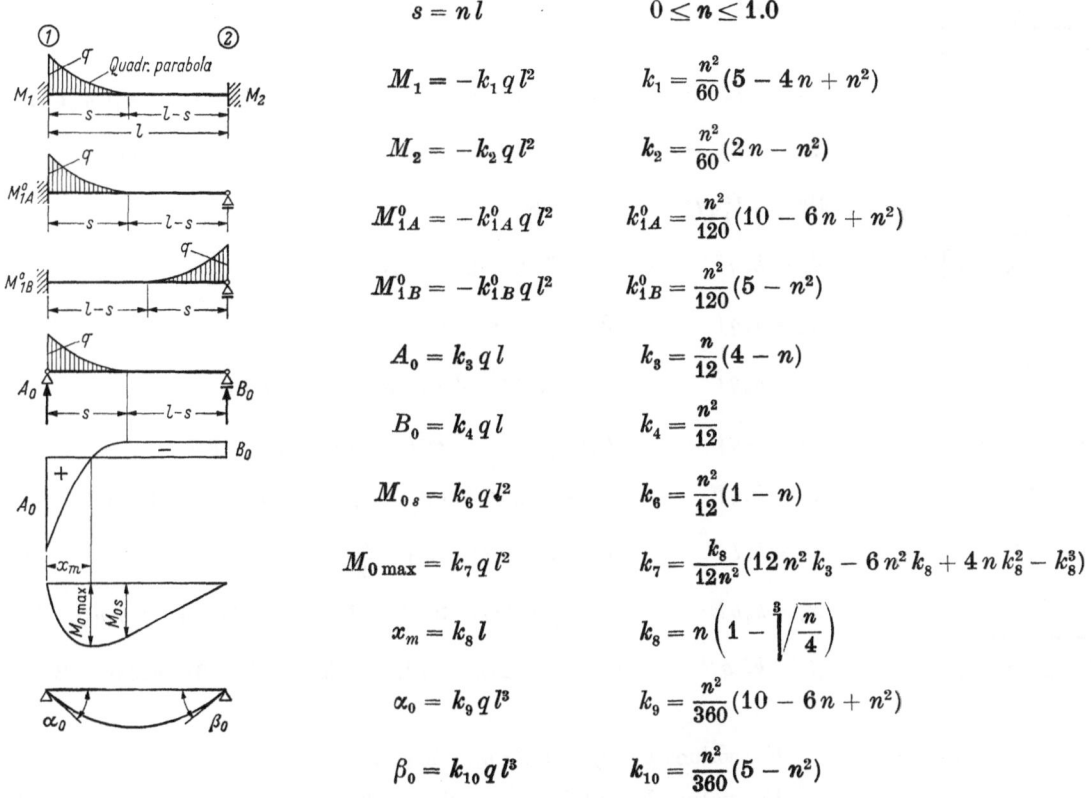

$$s = nl \qquad 0 \le n \le 1.0$$

$$M_1 = -k_1 q l^2 \qquad k_1 = \frac{n^2}{60}(5 - 4n + n^2)$$

$$M_2 = -k_2 q l^2 \qquad k_2 = \frac{n^2}{60}(2n - n^2)$$

$$M_{1A}^0 = -k_{1A}^0 q l^2 \qquad k_{1A}^0 = \frac{n^2}{120}(10 - 6n + n^2)$$

$$M_{1B}^0 = -k_{1B}^0 q l^2 \qquad k_{1B}^0 = \frac{n^2}{120}(5 - n^2)$$

$$A_0 = k_3 q l \qquad k_3 = \frac{n}{12}(4 - n)$$

$$B_0 = k_4 q l \qquad k_4 = \frac{n^2}{12}$$

$$M_{0s} = k_6 q l^2 \qquad k_6 = \frac{n^2}{12}(1 - n)$$

$$M_{0\,max} = k_7 q l^2 \qquad k_7 = \frac{k_8}{12 n^2}(12 n^2 k_3 - 6 n^2 k_8 + 4 n k_8^2 - k_8^3)$$

$$x_m = k_8 l \qquad k_8 = n \left(1 - \sqrt[3]{\frac{n}{4}}\right)$$

$$\alpha_0 = k_9 q l^3 \qquad k_9 = \frac{n^2}{360}(10 - 6n + n^2)$$

$$\beta_0 = k_{10} q l^3 \qquad k_{10} = \frac{n^2}{360}(5 - n^2)$$

Table 21

n	k_1	k_2	k_{1A}^0	k_{1B}^0	k_3	k_4	k_6	k_7	k_8	k_9	k_{10}	n
0	0	0	0	0	0	0	0	0	0	0	0	0
0.05	0.000	0.000	0.000	0.000	0.016	0.000	0.000	0.000	0.038	0.000	.000	0.05
0.1	.001	.000	.001	.000	.033	.001	.001	.001	.071	.000	0.00	0.1
0.15	.002	.000	.002	.001	.048	.002	.002	.002	.100	.001	.000	0.15
0.2	.003	.000	.003	.002	.063	.003	.003	.003	.126	.001	.001	0.2
0.25	.004	.000	.004	.003	.078	.005	.004	.004	.151	.001	.001	0.25
0.3	.006	.001	.006	.004	.093	.008	.005	.006	.173	.002	.001	0.3
0.35	.008	.001	.008	.005	.106	.010	.007	.008	.195	.003	.002	0.35
0.4	.009	.002	.010	.006	.120	.013	.008	.010	.214	.003	.002	0.4
0.45	.011	.002	.013	.008	.133	.017	.009	.012	.233	.004	.003	0.45
0.5	.014	.003	.015	.010	.146	.021	.010	.014	.250	.005	.003	0.5
0.55	.016	.004	.018	.012	.158	.025	.011	.017	.266	.006	.004	0.55
0.6	.018	.005	.020	.014	.170	.030	.012	.019	.281	.007	.005	0.6
0.65	.020	.006	.023	.016	.181	.035	.012	.022	.295	.008	.005	0.65
0.7	.022	.007	.026	.018	.193	.041	.012	.024	.308	.009	.006	0.7
0.75	.024	.009	.028	.021	.203	.047	.012	.027	.321	.009	.007	0.75
0.8	.026	.010	.031	.023	.213	.055	.011	.029	.332	.010	.008	0.8
0.85	.028	.012	.034	.026	.223	.060	.009	.032	.343	.011	.009	0.85
0.9	.030	.013	.037	.028	.233	.068	.007	.034	.353	.012	.009	0.9
0.95	.032	.015	.039	.031	.241	.075	.004	.037	.362	.013	.010	0.95
1.0	0.033	0.017	0.042	0.033	0.250	0.083	0.000	0.037	0.370	0.014	0.011	1.0

Table 22 A   – 54 –

Loading 22 (to Tables 22 A–E)

Free uniformly distributed line load

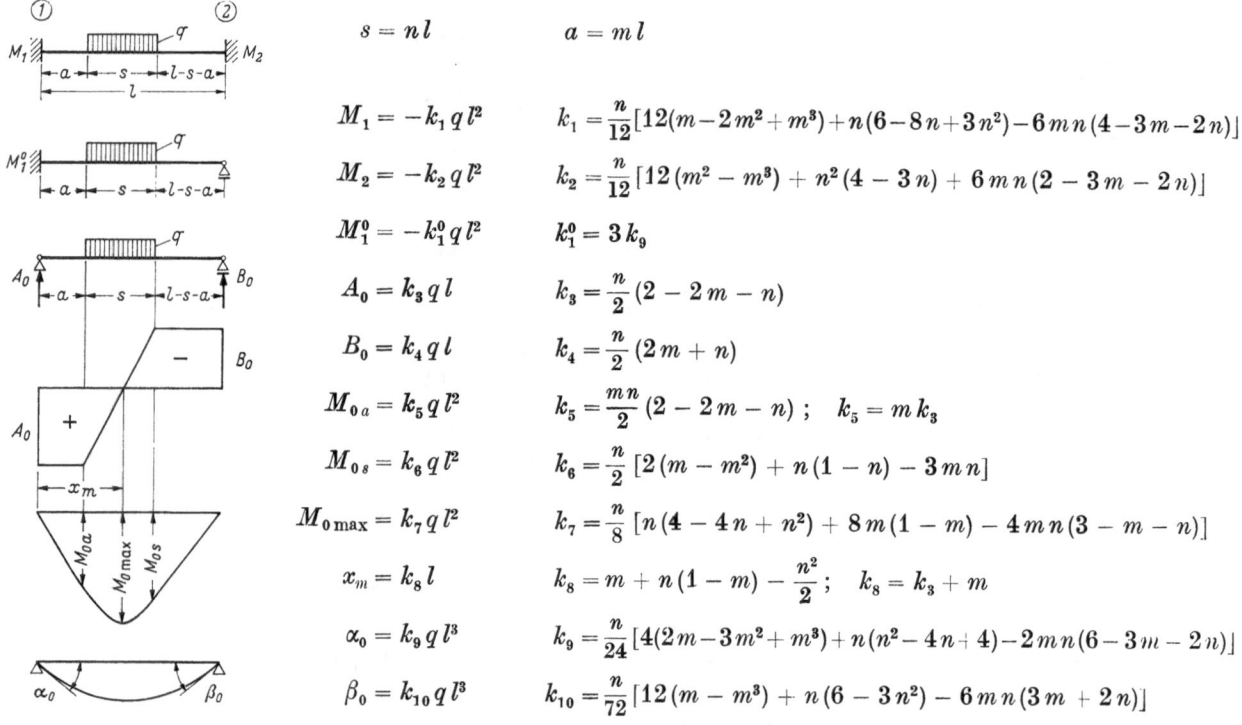

$$s = n\,l \qquad\qquad a = m\,l$$

$$M_1 = -k_1\,q\,l^2 \qquad k_1 = \frac{n}{12}\big[12(m - 2m^2 + m^3) + n(6 - 8n + 3n^2) - 6mn(4 - 3m - 2n)\big]$$

$$M_2 = -k_2\,q\,l^2 \qquad k_2 = \frac{n}{12}\big[12(m^2 - m^3) + n^2(4 - 3n) + 6mn(2 - 3m - 2n)\big]$$

$$M_1^0 = -k_1^0\,q\,l^2 \qquad k_1^0 = 3\,k_9$$

$$A_0 = k_3\,q\,l \qquad k_3 = \frac{n}{2}(2 - 2m - n)$$

$$B_0 = k_4\,q\,l \qquad k_4 = \frac{n}{2}(2m + n)$$

$$M_{0a} = k_5\,q\,l^2 \qquad k_5 = \frac{mn}{2}(2 - 2m - n)\,;\quad k_5 = m\,k_3$$

$$M_{0s} = k_6\,q\,l^2 \qquad k_6 = \frac{n}{2}\big[2(m - m^2) + n(1 - n) - 3mn\big]$$

$$M_{0\,\text{max}} = k_7\,q\,l^2 \qquad k_7 = \frac{n}{8}\big[n(4 - 4n + n^2) + 8m(1 - m) - 4mn(3 - m - n)\big]$$

$$x_m = k_8\,l \qquad k_8 = m + n(1 - m) - \frac{n^2}{2}\,;\quad k_8 = k_3 + m$$

$$\alpha_0 = k_9\,q\,l^3 \qquad k_9 = \frac{n}{24}\big[4(2m - 3m^2 + m^3) + n(n^2 - 4n + 4) - 2mn(6 - 3m - 2n)\big]$$

$$\beta_0 = k_{10}\,q\,l^3 \qquad k_{10} = \frac{n}{72}\big[12(m - m^3) + n(6 - 3n^2) - 6mn(3m + 2n)\big]$$

For values k_1, k_2 and k_1^0: see Table 22 A
For values k_3 and k_4:   see Table 22 B
For values k_5 and k_6:   see Table 22 C
For values k_7 and k_8:   see Table 22 D
For values k_9 and k_{10}:   see Table 22 E

Table 22 A

n		m										n
		0	0.1	0.2	0.3	0.4	0.5	0.6	0.7	0.8	0.9	
0	k_1 k_2 k_1^0	0	0	0	0	0	0	0	0	0	0	k_1 0 k_2 k_1^0
0.05	k_1 k_2 k_1^0	0.001 .000 .001	0.005 .001 .005	0.007 .002 .008	0.007 .004 .009	0.007 .005 .010	0.006 .007 .009	0.004 .007 .008	0.003 .007 .006	0.001 .006 .004	0.000 .003 .002	k_1 k_2 0.05 k_1^0
0.1	k_1 k_2 k_1^0	.004 .000 .005	.011 .002 .012	.014 .005 .016	.015 .008 .019	.014 .011 .019	.011 .014 .018	.008 .015 .015	.005 .014 .012	.002 .011 .007	.000 .004 .002	k_1 k_2 0.1 k_1^0
0.15	k_1 k_2 k_1^0	.009 .001 .010	.017 .004 .019	.022 .008 .026	.022 .013 .028	.019 .018 .028	.016 .021 .026	.011 .022 .022	.006 .020 .016	.002 .014 .009		k_1 k_2 0.15 k_1^0

Table 22 A (Continuation)

n		\(m\) = 0	0.1	0.2	0.3	0.4	0.5	0.6	0.7	0.8	0.9	n
0.2	k_1	0.015	0.025	0.029	0.028	0.025	0.019	0.013	0.007	0.002		k_1 0.2
	k_2	.002	.007	.013	.019	.025	.028	.029	.025	.015		k_2
	k_1^0	.016	.028	.035	.038	.037	.033	.027	.019	.010		k_1^0
0.25	k_1	.022	.032	.036	.034	.029	.022	.014	.007			k_1 0.25
	k_2	.004	.010	.018	.026	.032	.035	.035	.028			k_2
	k_1^0	.024	.037	.045	.047	.045	.040	.031	.021			k_1^0
0.3	k_1	.029	.039	.042	.039	.033	.024	.015	.007			k_1 0.3
	k_2	.007	.015	.024	.033	.039	.042	.039	.029			k_2
	k_1^0	.033	.047	.054	.056	.052	.045	.034	.021			k_1^0
0.35	k_1	.036	.046	.048	.044	.035	.025	.015				k_1 0.35
	k_2	.011	.020	.030	.040	.047	.048	.043				k_2
	k_1^0	.042	.056	.063	.064	.059	.049	.036				k_1^0
0.4	k_1	.044	.053	.053	.047	.037	.026	.015				k_1 0.4
	k_2	.015	.026	.037	.047	.053	.053	.044				k_2
	k_1^0	.051	.066	.072	.071	.064	.052	.037				k_1^0
0.45	k_1	.051	.059	.058	.050	.039	.026					k_1 0.45
	k_2	.020	.032	.045	.055	.059	.056					k_2
	k_1^0	.061	.075	.080	.077	.068	.054					k_1^0
0.5	k_1	.057	.064	.061	.052	.039	.026					k_1 0.5
	k_2	.026	.039	.052	.061	.064	.057					k_2
	k_1^0	.070	.084	.087	.083	.071	.055					k_1^0
0.55	k_1	.063	.068	.064	.053	.040						k_1 0.55
	k_2	.033	.047	.059	.067	.067						k_2
	k_1^0	.079	.092	.094	.087	.073						k_1^0
0.6	k_1	.068	.072	.066	.054	.040						k_1 0.6
	k_2	.040	.054	.066	.072	.068						k_2
	k_1^0	.088	.099	.099	.090	.074						k_1^0
0.65	k_1	.073	.075	.067	.054							k_1 0.65
	k_2	.047	.061	.072	.075							k_2
	k_1^0	.096	.105	.103	.092							k_1^0
0.7	k_1	.076	.077	.068	.054							k_1 0.7
	k_2	.054	.068	.077	.076							k_2
	k_1^0	.104	.111	.106	.092							k_1^0
0.75	k_1	.079	.078	.068								k_1 0.75
	k_2	.062	.074	.080								k_2
	k_1^0	.110	.115	.108								k_1^0
0.8	k_1	.081	.079	.068								k_1 0.8
	k_2	.068	.079	.081								k_2
	k_1^0	.115	.118	.109								k_1^0
0.85	k_1	.082	.079									k_1 0.85
	k_2	.074	.082									k_2
	k_1^0	.119	.120									k_1^0
0.9	k_1	.083	.079									k_1 0.9
	k_2	.079	.083									k_2
	k_1^0	.123	.121									k_1^0
0.95	k_1	.083										k_1 0.95
	k_2	.082										k_2
	k_1^0	.124										k_1^0
1.0	k_1	.083										k_1 1.0
	k_2	.083										k_2
	k_1^0	.125										k_1^0

Table 22 B   – 56 –

Table 22 B

n	m										n
	0	0.1	0.2	0.3	0.4	0.5	0.6	0.7	0.8	0.9	
0 $\begin{array}{l}k_3\\k_4\end{array}$	0	0	0	0	0	0	0	0	0	0	$\begin{array}{l}k_3\\k_4\end{array}$ 0
0.05 $\begin{array}{l}k_3\\k_4\end{array}$	0.049 .001	0.044 .006	0.039 .011	0.034 .016	0.029 .021	0.024 .026	0.019 .031	0.014 .036	0.009 .041	0.004 .046	$\begin{array}{l}k_3\\k_4\end{array}$ 0.05
0.1 $\begin{array}{l}k_3\\k_4\end{array}$	.095 .005	.085 .015	.075 .025	.065 .035	.055 .045	.045 .055	.035 .065	.025 .075	.015 .085	.005 .095	$\begin{array}{l}k_3\\k_4\end{array}$ 0.1
0.15 $\begin{array}{l}k_3\\k_4\end{array}$	.139 .011	.124 .026	.109 .041	.094 .056	.079 .071	.064 .086	.049 .101	.034 .116	.019 .131		$\begin{array}{l}k_3\\k_4\end{array}$ 0.15
0.2 $\begin{array}{l}k_3\\k_4\end{array}$	.180 .020	.160 .040	.140 .060	.120 .080	.100 .100	.080 .120	.060 .140	.040 .160	.020 .180		$\begin{array}{l}k_3\\k_4\end{array}$ 0.2
0.25 $\begin{array}{l}k_3\\k_4\end{array}$	.219 .031	.194 .056	.169 .081	.144 .106	.119 .131	.094 .156	.069 .181	.044 .206			$\begin{array}{l}k_3\\k_4\end{array}$ 0.25
0.3 $\begin{array}{l}k_3\\k_4\end{array}$	.255 .045	.225 .075	.195 .105	.165 .135	.135 .165	.105 .195	.075 .225	.045 .255			$\begin{array}{l}k_3\\k_4\end{array}$ 0.3
0.35 $\begin{array}{l}k_3\\k_4\end{array}$	.289 .061	.254 .096	.219 .131	.184 .166	.149 .201	.114 .236	.079 .271				$\begin{array}{l}k_3\\k_4\end{array}$ 0.35
0.4 $\begin{array}{l}k_3\\k_4\end{array}$	.320 .080	.280 .120	.240 .160	.200 .200	.160 .240	.120 .280	.080 .320				$\begin{array}{l}k_3\\k_4\end{array}$ 0.4
0.45 $\begin{array}{l}k_3\\k_4\end{array}$	.349 .101	.304 .146	.259 .191	.214 .236	.169 .281	.124 .326					$\begin{array}{l}k_3\\k_4\end{array}$ 0.45
0.5 $\begin{array}{l}k_3\\k_4\end{array}$	.375 .125	.325 .175	.275 .225	.225 .275	.175 .325	.125 .375					$\begin{array}{l}k_3\\k_4\end{array}$ 0.5
0.55 $\begin{array}{l}k_3\\k_4\end{array}$	.399 .151	.344 .206	.289 .261	.234 .316	.179 .371						$\begin{array}{l}k_3\\k_4\end{array}$ 0.55
0.6 $\begin{array}{l}k_3\\k_4\end{array}$	.420 .180	.360 .240	.300 .300	.240 .360	.180 .420						$\begin{array}{l}k_3\\k_4\end{array}$ 0.6
0.65 $\begin{array}{l}k_3\\k_4\end{array}$	.439 .211	.374 .276	.309 .341	.244 .406							$\begin{array}{l}k_3\\k_4\end{array}$ 0.65
0.7 $\begin{array}{l}k_3\\k_4\end{array}$	.455 .245	.385 .315	.315 .385	.245 .455							$\begin{array}{l}k_3\\k_4\end{array}$ 0.7
0.75 $\begin{array}{l}k_3\\k_4\end{array}$	.469 .281	.394 .356	.319 .431								$\begin{array}{l}k_3\\k_4\end{array}$ 0.75
0.8 $\begin{array}{l}k_3\\k_4\end{array}$	.480 .320	.400 .400	.320 .480								$\begin{array}{l}k_3\\k_4\end{array}$ 0.8
0.85 $\begin{array}{l}k_3\\k_4\end{array}$	.489 .361	.404 .446									$\begin{array}{l}k_3\\k_4\end{array}$ 0.85
0.9 $\begin{array}{l}k_3\\k_4\end{array}$	.495 .405	.405 .495									$\begin{array}{l}k_3\\k_4\end{array}$ 0.9
0.95 $\begin{array}{l}k_3\\k_4\end{array}$	.499 .451										$\begin{array}{l}k_3\\k_4\end{array}$ 0.95
1.0 $\begin{array}{l}k_3\\k_4\end{array}$	.500 .500										$\begin{array}{l}k_3\\k_4\end{array}$ 1.0

Table 22C

n		m										n	
		0	0.1	0.2	0.3	0.4	0.5	0.6	0.7	0.8	0.9		
0	k_5	0	0	0	0	0	0	0	0	0	0	k_5	0
	k_6											k_6	
0.05	k_5	0	0.004	0.008	0.010	0.012	0.012	0.011	0.010	0.007	0.003	k_5	0.05
	k_6	0.001	.005	.008	.010	.012	.012	.011	.009	.006	.002	k_6	
0.1	k_5	0	.009	.015	.020	.022	.023	.021	.018	.012	.005	k_5	0.1
	k_6	.005	.012	.018	.021	.023	.022	.020	.015	.009	.000	k_6	
0.15	k_5	0	.012	.022	.028	.032	.032	.029	.024	.015		k_5	0.15
	k_6	.010	.020	.027	.031	.032	.030	.025	.017	.007		k_6	
0.2	k_5	0	.016	.028	.036	.040	.040	.036	.028	.016		k_5	0.2
	k_6	.016	.028	.036	.040	.040	.036	.028	.016	.000		k_6	
0.25	k_5	0	.019	.034	.043	.048	.047	.041	.031			k_5	0.25
	k_6	.023	.037	.045	.048	.046	.039	.027	.010			k_6	
0.3	k_5	0	.023	.039	.050	.054	.053	.045	.032			k_5	0.3
	k_6	.032	.045	.053	.054	.050	.039	.023	.000			k_6	
0.35	k_5	0	.025	.044	.055	.060	.057	.047				k_5	0.35
	k_6	.040	.053	.059	.058	.050	.035	.014				k_6	
0.4	k_5	0	.028	.048	.060	.064	.060	.048				k_5	0.4
	k_6	.048	.060	.064	.060	.048	.028	.000				k_6	
0.45	k_5	0	.030	.052	.064	.068	.062					k_5	0.45
	k_6	.056	.066	.067	.059	.042	.016					k_6	
0.5	k_5	0	.033	.055	.068	.070	.063					k_5	0.5
	k_6	.063	.070	.068	.055	.033	.000					k_6	
0.55	k_5	0	.034	.058	.070	.072						k_5	0.55
	k_6	.068	.072	.065	.047	.019						k_6	
0.6	k_5	0	.036	.060	.072	.072						k_5	0.6
	k_6	.072	.072	.060	.036	.000						k_6	
0.65	k_5	0	.037	.062	.073							k_5	0.65
	k_6	.074	.069	.051	.020							k_6	
0.7	k_5	0	.039	.063	.074							k_5	0.7
	k_6	.074	.063	.039	.000							k_6	
0.75	k_5	0	.039	.064								k_5	0.75
	k_6	.070	.053	.022								k_6	
0.8	k_5	0	.040	.064								k_5	0.8
	k_6	.064	.040	.000								k_6	
0.85	k_5	0	.040									k_5	0.85
	k_6	.054	.022									k_6	
0.9	k_5	0	.041									k_5	0.9
	k_6	.041	.000									k_6	
0.95	k_5	0										k_5	0.95
	k_6	.023										k_6	
1.0	k_5	0										k_5	1.0
	k_6	.000										k_6	

Table 22 D   – 58 –

Table 22 D

n		0	0.1	0.2	0.3	0.4	0.5	0.6	0.7	0.8	0.9		n
							m						
0	k_7	0	0	0	0	0	0	0	0	0	0	k_7	0
	k_8											k_8	
0.05	k_7	0.001	0.005	0.009	0.011	0.012	0.012	0.011	0.010	0.007	0.003	k_7	0.05
	k_8	.049	.144	.239	.334	.429	.524	.619	.714	.809	.904	k_8	
0.1	k_7	.005	.012	.018	.022	.024	.024	.022	.018	.012	.005	k_7	0.1
	k_8	.095	.185	.275	.365	.455	.545	.635	.725	.815	.905	k_8	
0.15	k_7	.010	.020	.028	.033	.035	.034	.030	.024	.015		k_7	0.15
	k_8	.139	.224	.309	.394	.479	.564	.649	.734	.819		k_8	
0.2	k_7	.016	.029	.038	.043	.045	.043	.038	.029	.016		k_7	0.2
	k_8	.180	.260	.340	.420	.500	.580	.660	.740	.820		k_8	
0.25	k_7	.024	.038	.048	.053	.055	.051	.043	.032			k_7	0.25
	k_8	.219	.294	.369	.444	.519	.594	.669	.744			k_8	
0.3	k_7	.033	.048	.058	.063	.063	.058	.048	.033			k_7	0.3
	k_8	.255	.325	.395	.465	.535	.605	.675	.745			k_8	
0.35	k_7	.042	.058	.068	.072	.071	.063	.050				k_7	0.35
	k_8	.289	.354	.419	.484	.549	.614	.679				k_8	
0.4	k_7	.051	.067	.077	.080	.077	.067	.051				k_7	0.4
	k_8	.320	.380	.440	.500	.560	.620	.680				k_8	
0.45	k_7	.061	.077	.085	.087	.082	.070					k_7	0.45
	k_8	.349	.404	.459	.514	.569	.624					k_8	
0.5	k_7	.070	.085	.093	.093	.085	.070					k_7	0.5
	k_8	.375	.425	.475	.525	.575	.625					k_8	
0.55	k_7	.080	.093	.099	.097	.087						k_7	0.55
	k_8	.399	.444	.489	.534	.579						k_8	
0.6	k_7	.088	.101	.105	.101	.088						k_7	0.6
	k_8	.420	.460	.500	.540	.580						k_8	
0.65	k_7	.096	.107	.109	.103							k_7	0.65
	k_8	.439	.474	.509	.544							k_8	
0.7	k_7	.104	.113	.113	.104							k_8	0.7
	k_8	.455	.485	.515	.545							k_7	
0,75	k_7	.110	.117	.114								k_7	0.75
	k_8	.469	.494	.519								k_8	
0.8	k_7	.115	.120	.115								k_7	0.8
	k_8	.480	.500	.520								k_8	
0.85	k_7	.119	.122									k_7	0.85
	k_8	.489	.504									k_8	
0.9	k_7	.123	.123									k_7	0.9
	k_8	.495	.505									k_8	
0.95	k_7	.124										k_7	0.95
	k_8	.499										k_8	
1.0	k_7	.125										k_7	1.0
	k_8	.500										k_8	

Table 22 E

n		0	0.1	0.2	0.3	0.4	0.5	0.6	0.7	0.8	0.9		n
						m							
0	k_9	0	0	0	0	0	0	0	0	0	0	k_9	0
	k_{10}											k_{10}	
0.05	k_9	0.000	0.002	0.003	0.003	0.003	0.003	0.003	0.002	0.001	0.001	k_9	0.05
	k_{10}	.000	.001	.002	.002	.003	.003	.003	.003	.002	.001	k_{10}	
0.1	k_9	.002	.004	.005	.006	.006	.006	.005	.004	.002	.001	k_9	0.1
	k_{10}	.001	.002	.004	.005	.006	.006	.006	.005	.004	.002	k_{10}	
0.15	k_9	.003	.006	.009	.009	.009	.009	.007	.005	.003		k_9	0.15
	k_{10}	.002	.004	.006	.008	.009	.010	.009	.008	.005		k_{10}	
0.2	k_9	.005	.009	.012	.013	.012	.011	.009	.006	.003		k_9	0.2
	k_{10}	.003	.006	.009	.011	.012	.013	.012	.009	.005		k_{10}	
0.25	k_9	.008	.012	.015	.016	.015	.013	.010	.007			k_9	0.25
	k_{10}	.005	.009	.012	.014	.016	.015	.014	.010			k_{10}	
0.3	k_9	.011	.016	.018	.019	.017	.015	.011	.007			k_9	0.3
	k_{10}	.007	.011	.015	.017	.019	.018	.016	.011			k_{10}	
0.35	k_9	.014	.019	.021	.021	.020	.016	.012				k_9	0.35
	k_{10}	.010	.014	.018	.021	.021	.020	.017				k_{10}	
0.4	k_9	.017	.022	.024	.024	.021	.017	.012				k_9	0.4
	k_{10}	.012	.017	.021	.024	.024	.022	.017				k_{10}	
0.45	k_9	.020	.025	.027	.026	.023	.018					k_9	0.45
	k_{10}	.015	.021	.025	.027	.026	.023					k_{10}	
0.5	k_9	.023	.028	.029	.028	.024	.018					k_9	0.5
	k_{10}	.018	.024	.028	.029	.028	.023					k_{10}	
0.55	k_9	.027	.031	.031	.029	.024						k_9	0.55
	k_{10}	.021	.027	.030	.031	.029						k_{10}	
0.6	k_9	.029	.033	.033	.030	.025						k_9	0.6
	k_{10}	.025	.030	.033	.033	.029						k_{10}	
0.65	k_9	.032	.035	.034	.031							k_9	0.65
	k_{10}	.028	.033	.035	.034							k_{10}	
0.7	k_9	.035	.037	.035	.031							k_9	0.7
	k_{10}	.031	.035	.037	.035							k_{10}	
0.75	k_9	.037	.038	.036								k_9	0.75
	k_{10}	.034	.038	.038								k_{10}	
0.8	k_9	.038	.039	.036								k_9	0.8
	k_{10}	.036	.039	.038								k_{10}	
0.85	k_9	.040	.040									k_9	0.85
	k_{10}	.038	.040									k_{10}	
0.9	k_9	.041	.040									k_9	0.9
	k_{10}	.040	.041									k_{10}	
0.95	k_9	.041										k_9	0.95
	k_{10}	.041										k_{10}	
1.0	k_9	.042										k_9	1.0
	k_{10}	.042										k_{10}	

Loading 23 (to Tables 23 A–F)

Free unsymmetrical triangular line load

$$s = n\,l$$
$$a = m\,l$$
$$0 \leq (n + m) \leq 1.0$$

$$M_1 = -k_1 q\,l^2 \qquad k_1 = \frac{n}{60}\left[30\,m(m^2 - 2m + 1) + n(3n^2 - 10n + 10) + \right.$$
$$\left. + 5\,m\,n(6m - 8 + 3n)\right]$$

$$M_2 = -k_2 q\,l^2 \qquad k_2 = \frac{n}{60}\left[30\,m^2(1 - m) + n^2(5 - 3n) + \right.$$
$$\left. + 5\,m\,n(4 - 6m - 3n)\right]$$

$$M_{1A}^0 = -k_{1A}^0 q\,l^2 \quad k_{1A}^0 = \frac{n}{120}\left[30\,m(2 - 3m + m^2) + n(20 - 15n + 3n^2) + \right.$$
$$\left. + 15\,m\,n(2m - 4 + n)\right]$$

$$M_{1B}^0 = -k_{1B}^0 q\,l^2 \quad k_{1B}^0 = \frac{n}{120}\left[30\,m(2 - 3m + m^2) + n(40 - 45n + 12n^2) - \right.$$
$$\left. - 15\,m\,n(8 - 4m - 3n)\right]$$

$$A_0 = k_3 q\,l \qquad k_3 = \frac{n}{6}(3 - 3m - n)$$

$$B_0 = k_4 q\,l \qquad k_4 = \frac{n}{6}(3m + n)$$

$$M_{0a} = k_5 q\,l^2 \qquad k_5 = \frac{m\,n}{6}(3 - 3m - n); \quad k_5 = m\,k_3$$

$$M_{0s} = k_6 q\,l^2 \qquad k_6 = \frac{n}{6}(3m + n)(1 - m - n); \quad k_6 = k_4(1 - m - n)$$

$$M_{0\max} = k_7 q\,l^2 \qquad k_7 = \frac{1}{6n}\left[3\,k_8(2n\,k_3 + 2m\,n + m^2) - 3\,k_8^2(m + n) + \right.$$
$$\left. + k_8^3 - m^2(3n + m)\right]$$

$$x_m = k_8 l \qquad k_8 = m + n\left(1 - 0.57733\,\sqrt{3m + n}\right)$$

$$\alpha_0 = k_9 q\,l^3 \qquad k_9 = \frac{n}{360}\left[30\,m(2 - 3m + m^2) + n(20 - 15n + 3n^2) + \right.$$
$$\left. + 15\,m\,n(2m - 4 + n)\right]$$

$$\beta_0 = k_{10} q\,l^3 \qquad k_{10} = \frac{n}{360}\left[30\,m(1 - m^2) + n(10 - 3n^2) - \right.$$
$$\left. - 5\,m\,n(6m + 3n)\right]$$

In this Table k_{1B}^0 corresponds to an independent case of loading, the relations between the coefficients developed in the annexe not being suitable here for k_{1B}^0.

For values k_1 and k_2: see Table 23 A

For values k_{1A}^0 and k_{1B}^0: see Table 23 B

For values k_3 and k_4: see Table 23 C

For values k_5 and k_6: see Table 23 D

For values k_7 and k_8: see Table 23 E

For values k_9 and k_{10}: see Table 23 F

Table 23A

n		m										n	
		0	0.1	0.2	0.3	0.4	0.5	0.6	0.7	0.8	0.9		
0	k_1	0	0	0	0	0	0	0	0	0	0	k_1	0
	k_2											k_2	
0.05	k_1	0.000	0.002	0.003	0.004	0.004	0.003	0.002	0.001	0.001	0.000	k_1	0.05
	k_2	.000	.000	.001	.002	.003	.003	.004	.004	.003	.002	k_2	
0.1	k_1	.002	.005	.007	.007	.007	.006	.004	.003	.001	.000	k_1	0.1
	k_2	.000	.001	.002	.004	.005	.007	.007	.007	.006	.003	k_2	
0.15	k_1	.003	.008	.010	.011	.010	.008	.006	.004	.001		k_1	0.15
	k_2	.000	.001	.004	.006	.008	.010	.011	.010	.008		k_2	
0.2	k_1	.005	.011	.014	.015	.013	.011	.007	.004	.002		k_1	0.2
	k_2	.001	.002	.005	.008	.012	.014	.015	.013	.019		k_2	
0.25	k_1	.008	.015	.018	.018	.016	.013	.009	.005			k_1	0.25
	k_2	.001	.004	.007	.011	.015	.017	.018	.016			k_2	
0.3	k_1	.011	.018	.021	.021	.018	.014	.010	.005			k_1	0.3
	k_2	.002	.005	.010	.014	.018	.021	.021	.018			k_2	
0.35	k_1	.014	.022	.025	.024	.021	.016	.010				k_1	0.35
	k_2	.003	.007	.012	.017	.022	.024	.024				k_2	
0.4	k_1	.017	.025	.028	.027	.023	.017	.011				k_1	0.4
	k_2	.004	.009	.015	.021	.025	.028	.026				k_2	
0.45	k_1	.021	.029	.031	.029	.024	.018					k_1	0.45
	k_2	.006	.011	.018	.024	.029	.031					k_2	
0.5	k_1	.024	.032	.034	.031	.026	.019					k_1	
	k_2	.007	.014	.021	.027	.032	.033					k_2	0.5
0.55	k_1	.027	.035	.037	.033	.027						k_1	0.55
	k_2	.009	.016	.024	.031	.035						k_2	
0.6	k_1	.030	.038	.039	.035	.028						k_1	0.6
	k_2	.012	.019	.027	.034	.038						k_2	
0.65	k_1	.034	.041	.041	.036							k_1	0.65
	k_2	.014	.022	.030	.037							k_2	
0.7	k_1	.037	.043	.043	.038							k_1	0.7
	k_2	.017	.025	.034	.040							k_2	
0.75	k_1	.039	.045	.045								k_1	0.75
	k_2	.019	.028	.036								k_2	
0.8	k_1	.042	.048	.046								k_1	0.8
	k_2	.022	.031	.039								k_2	
0.85	k_1	.044	.049									k_1	0.85
	k_2	.025	.034									k_2	
0.9	k_1	0.46	.051									k_1	0.9
	k_2	0.28	.037									k_2	
0.95	k_1	.048										k_1	0.95
	k_2	.031										k_2	
1.0	k_1	.050										k_1	1.0
	k_2	.033										k_2	

Table 23 B   – 62 –

Table 23 B

n	m										n
	0	0.1	0.2	0.3	0.4	0.5	0.6	0.7	0.8	0.9	
0	0	0	0	0	0	0	0	0	0	0	0
0.05 $k\varrho_A$	0.000	0.002	0.004	0.005	0.005	0.005	0.004	0.003	0.002	0.001	$k\varrho_A$ 0.05
$k\varrho_B$	.001	.003	.004	.005	.005	.005	.004	.003	.002	.001	$k\varrho_B$
0.1 $k\varrho_A$	.002	.005	.008	.009	.010	.009	.008	.006	.004	.002	$k\varrho_A$ 0.1
$k\varrho_B$	.003	.006	.008	.009	.009	.009	.007	.005	.003	.001	$k\varrho_B$
0.15 $k\varrho_A$	.003	.009	.012	.014	.014	.013	.012	.009	.005		$k\varrho_A$ 0.15
$k\varrho_B$	.006	.011	.013	.014	.014	.013	.010	.007	.004		$k\varrho_B$
0.2 $k\varrho_A$	.006	.012	.017	.019	.019	.017	.015	.011	.006		$k\varrho_A$ 0.2
$k\varrho_B$	.010	.016	.018	.019	.018	.016	.012	.008	.003		$k\varrho_B$
0.25 $k\varrho_A$	.009	.016	.021	.024	.023	.021	.018	.013			$k\varrho_A$ 0.25
$k\varrho_B$	.015	.021	.023	.023	.022	.018	.014	.008			$k\varrho_A$
0.3 $k\varrho_A$	.012	.021	.026	.028	.028	.025	.020	.014			$k\varrho_A$ 0.3
$k\varrho_B$	.021	.026	.028	.028	.025	.020	.014	.007			$k\varrho_B$
0.35 $k\varrho_A$	.015	.025	.031	.033	.032	.028	.022				$k\varrho_A$ 0.35
$k\varrho_B$	.026	.031	.033	.031	.027	.021	.014				$k\varrho_B$
0.4 $k\varrho_A$	.019	.030	.035	.037	0.35	.031	.024				$k\varrho_A$ 0.4
$k\varrho_B$	.032	.036	.037	.034	0.29	.021	.013				$k\varrho_B$
0.45 $k\varrho_A$	.023	.034	.040	.041	.039	.033					$k\varrho_A$ 0.45
$k\varrho_B$	.037	.041	.040	.036	.030	.021					$k\varrho_B$
0.5 $k\varrho_A$	.028	.039	.044	.045	.042	.035					$k\varrho_A$ 0.5
$k\varrho_B$	.043	.045	.043	.038	.029	.019					$k\varrho_B$
0.55 $k\varrho_A$	.032	.043	.048	.049	.045						$k\varrho_A$ 0.55
$k\varrho_B$	.048	.049	.045	.038	.029						$k\varrho_B$
0.6 $k\varrho_A$	.036	.047	.052	.052	.047						$k\varrho_A$ 0.6
$k\varrho_B$	.052	.052	.047	.038	.027						$k\varrho_B$
0.65 $k\varrho_A$	.041	.052	.056	.055							$k\varrho_A$ 0.65
$k\varrho_B$	.056	.054	.047	.037							$k\varrho_B$
0.7 $k\varrho_A$	.045	.056	.060	.058							$k\varrho_A$ 0.7
$k\varrho_B$	.059	.055	.047	.035							$k\varrho_B$
0.75 $k\varrho_A$	.049	.060	.063								$k\varrho_A$ 0.75
$k\varrho_B$	.061	.055	.045								$k\varrho_B$
0.8 $k\varrho_A$	.053	.063	.066								$k\varrho_A$ 0.8
$k\varrho_B$	.062	.055	.043								$k\varrho_B$
0.85 $k\varrho_A$	.057	.066									$k\varrho_A$ 0.85
$k\varrho_B$	0.63	.053									$k\varrho_B$
0.9 $k\varrho_A$	.060	.069									$k\varrho_A$ 0.9
$k\varrho_B$	.062	.051									$k\varrho_B$
0.95 $k\varrho_A$	.064										$k\varrho_A$ 0.95
$k\varrho_B$	.061										$k\varrho_B$
1.0 $k\varrho_A$	.067										$k\varrho_A$ 0.1
$k\varrho_B$	.058										$k\varrho_B$

Table 23 B

Table 23 C

n	m										n
	0	0.1	0.2	0.3	0.4	0.5	0.6	0.7	0.8	0.9	
0	0	0	0	0	0	0	0	0	0	0	0
0.05 k_3	0.025	0.022	0.020	0.017	0.015	0.012	0.010	0.007	0.005	0.002	k_3 0.05
k_4	.000	.003	.005	.008	.010	.013	.015	.018	.020	.023	k_4
0.1 k_3	.048	.043	.038	.033	.028	.023	.018	.013	.008	.003	k_3 0.1
k_4	.002	.007	.012	.017	.022	.027	.032	.037	.042	.047	k_4
0.15 k_3	.071	.064	.056	.049	.041	.034	.026	.019	.011		k_3 0.15
k_4	.004	.011	.019	.026	.034	.041	.049	.056	.064		k_4
0.2 k_3	.093	.083	.073	.063	.053	.043	.033	.023	.013		k_3 0.2
k_4	.007	.017	.027	.037	.047	.057	.067	.077	.087		k_4
0.25 k_3	.115	.102	.090	.077	.065	.052	.040	.027			k_3 0.25
k_4	.010	.023	.035	.048	.060	.073	.085	.098			k_4
0.3 k_3	.135	.120	.105	.090	.075	.060	.045	.030			k_3 0.3
k_4	.015	.030	.045	.060	.075	.090	.105	.120			k_4
0.35 k_3	.155	.137	.120	.102	.085	.067	.050				k_3 0.35
k_4	.020	.038	.055	.073	.090	.108	.125				k_4
0.4 k_3	.173	.153	.133	.113	.093	.073	.053				k_3 0.4
k_4	.027	.047	.067	.087	.107	.127	.147				k_4
0.45 k_3	.191	.169	.146	.124	.101	.079					k_3 0.45
k_4	.034	.056	.079	.101	.124	.146					k_4
0.5 k_3	.208	.183	.158	.133	.108	.083					k_3 0.5
k_4	.042	.067	.092	.117	.142	.167					k_4
0.55 k_3	.225	.197	.170	.142	.115						k_3 0.55
k_4	.050	.078	.105	.133	.160						k_4
0.6 k_3	.240	.210	.180	.150	.120						k_3 0.6
k_4	.060	.090	.120	.150	.180						k_4
0.65 k_3	.255	.222	.190	.157							k_3 0.65
k_4	.070	.103	.135	.168							k_4
0.7 k_3	.268	.233	.198	.163							k_3 0.7
k_4	.082	.117	.152	.187							k_4
0.75 k_3	.281	.244	.206								k_3 0.75
k_4	.094	.131	.169								k_4
0.8 k_3	.293	.253	.213								k_3 0.8
k_4	.107	.147	.187								k_4
0.85 k_3	.305	.262									k_3 0.85
k_4	.120	.163									k_4
0.9 k_3	.315	.270									k_3 0.9
k_4	.135	.180									k_4
0.95 k_3	.325										k_3 0.95
k_4	.150										k_4
1.0 k_3	.333										k_3 1.0
k_4	.167										k_4

Table 23 D — 64 —

Table 23 D

n		0	0.1	0.2	0.3	0.4	0.5	0.6	0.7	0.8	0.9	n	
						m								
0	k_5	0	0	0	0	0	0	0	0	0	0	k_5	0
	k_6											k_6	
0.05	k_5	0	0.002	0.004	0.005	0.006	0.006	0.006	0.006	0.004	0.002	k_5	0.05
	k_6	0.000	.002	.004	.005	.006	.006	.005	.004	.003	.001	k_6	
0.1	k_5	0	.004	.008	.010	.011	.012	.011	.009	.007	.003	k_5	0.1
	k_6	.002	.005	.008	.010	.011	.011	.010	.007	.004	0	k_6	
0.15	k_5	0	.006	.011	.015	.017	.017	.016	.013	.009		k_5	0.15
	k_6	.003	.008	.012	.014	.015	.014	.012	.008	.003		k_6	
0.2	k_5	0	.008	.015	.019	.021	.022	.020	.016	.011		k_5	0.2
	k_6	.005	.012	.016	.018	.019	.017	.013	.008	0		k_6	
0.25	k_5	0	.010	.018	.023	.026	.026	.024	.019			k_5	0.25
	k_6	.008	.015	.019	.022	.021	.018	.013	.005			k_6	
0.3	k_5	0	.012	.021	.027	.030	.030	.027	.021			k_5	0.3
	k_6	.011	.018	.023	.024	.023	.018	.011	0			k_6	
0.35	k_5	0	.014	.024	.031	.034	.034	.030				k_5	0.35
	k_6	.013	.021	.025	.026	.023	.016	.006				k_6	
0.4	k_5	0	.015	.027	.034	.037	.037	.032				k_5	0.4
	k_6	.016	.023	.027	.026	.021	.013	0				k_6	
0.45	k_5	0	.017	.029	.037	.041	.039					k_5	0.45
	k_6	.019	.025	.028	.025	.019	.007					k_6	
0.5	k_5	0	.018	.032	.040	.043	.042					k_5	0.5
	k_6	.021	.027	.028	.023	.014	0					k_6	
0.55	k_5	0	.020	.034	.043	.046						k_5	0.55
	k_6	.023	.027	.026	.020	.008						k_6	
0.6	k_5	0	.021	.036	.045	.048						k_5	0.6
	k_6	.024	.027	.024	.015	0						k_6	
0.65	k_5	0	.022	.038	.047							k_5	0.65
	k_6	.025	.026	.020	.008							k_6	
0.7	k_5	0	.023	.040	.049							k_5	0.7
	k_6	.025	.023	.015	0							k_6	
0.75	k_5	0	.024	.041								k_5	0.75
	k_6	.023	.020	.008								k_6	
0.8	k_5	0	.025	.043								k_5	0.8
	k_6	.021	.015	0								k_6	
0.85	k_5	0	.026									k_5	0.85
	k_6	.018	.008									k_6	
0.9	k_5	0	.027									k_5	0.9
	k_6	.014	0									k_6	
0.95	k_5	0										k_5	0.95
	k_6	.008										k_6	
1.0	k_5	0										k_5	1.0
	k_6	0										k_6	

Table 23 E

n		m										n	
		0	0.1	0.2	0.3	0.4	0.5	0.6	0.7	0.8	0.9		
0	k_7	0	0	0	0	0	0	0	0	0	0	k_7	0
	k_8	0	0.100	0.200	0.300	0.400	0.500	0.600	0.700	0.800	0.900	k_8	
0.05	k_7	0.000	.003	.004	.005	.006	.006	.006	.005	.004	.002	k_7	0.05
	k_8	.043	.133	.227	.322	.418	.514	.611	.708	.805	.902	k_8	
0.1	k_7	.002	.005	.009	.011	.012	.012	.011	.009	.007	.003	k_7	0.1
	k_8	.082	.163	.252	.342	.434	.527	.620	.714	.809	.903	k_8	
0.15	k_7	.003	.009	.013	.016	.017	.018	.016	.013	.009		k_7	0.15
	k_8	.116	.192	.275	.361	.449	.539	.629	.720	.812		k_8	
0.2	k_7	.006	.013	.018	.021	.024	.023	.021	.017	.011		k_7	0.2
	k_8	.148	.218	.297	.379	.463	.549	.637	.725	.814		k_8	
0.25	k_7	.008	.017	.023	.026	.028	.028	.025	.019			k_7	0.25
	k_8	.178	.243	.317	.395	.476	.559	.643	.729			k_8	
0.3	k_7	.011	.021	.027	.032	.033	.032	.028	.021			k_7	0.3
	k_8	.205	.266	.336	.410	.488	.568	.649	.732			k_8	
0.35	k_7	.015	.025	.032	.037	.038	.036	.031				k_7	0.35
	k_8	.230	.287	.353	.424	.498	.575	.654				k_8	
0.4	k_7	.019	.029	.037	.041	.042	.040	.033				k_7	0.4
	k_8	.254	.307	.369	.437	.508	.582	.657				k_8	
0.45	k_7	.022	.034	.042	.046	.046	.043					k_7	0.45
	k_8	.276	.325	.384	.448	.516	.587					k_8	
0.5	k_7	.027	.038	.046	.050	.050	.045					k_7	0.5
	k_8	.296	.342	.397	.458	.524	.592					k_8	
0.55	k_7	.031	.042	.050	.054	.053						k_7	0.55
	k_8	.315	.357	.409	.468	.530						k_8	
0.6	k_7	.035	.047	.054	.057	.056						k_7	0.6
	k_8	.332	.371	.421	.476	.535						k_8	
0.65	k_7	.039	.051	.058	.061							k_7	0.65
	k_8	.347	.384	.430	.483							k_8	
0.7	k_7	.043	.055	.062	.064							k_7	0.7
	k_8	.362	.396	.439	.489							k_8	
0.75	k_7	.047	.059	.065								k_7	0.75
	k_8	.375	.406	.447								k_8	
0.8	k_7	.051	.062	.068								k_7	0,8
	k_8	.387	.416	.454								k_8	
0.85	k_7	.054	.065									k_7	0.85
	k_8	.398	.424									k_8	
0.9	k_7	.058	.068									k_7	0.9
	k_8	.407	.431									k_8	
0.95	k_7	.061										k_7	0.95
	k_8	.415										k_8	
1.0	k_7	.064										k_7	1.0
	k_8	.423										k_8	

Table 23 E

Table 23 F — 66 —

Table 23 F

n		m										n
		0	0.1	0.2	0.3	0.4	0.5	0.6	0.7	0.8	0.9	
0	k_9	0	0	0	0	0	0	0	0	0	0	k_9 0
	k_{10}	0	0	0	0	0	0	0	0	0	0	k_{10}
0.05	k_9	0.000	0.001	0.001	0.002	0.002	0.002	0.001	0.001	0.001	0.000	k_9 0.05
	k_{10}	.000	.000	.001	.001	.001	.002	.002	.001	.001	.001	k_{10}
0.1	k_9	.001	.002	.003	.003	.003	.003	.003	.002	.001	.001	k_9 0.1
	k_{10}	.000	.001	.002	.002	.003	.003	.003	.003	.002	.001	k_{10}
0.15	k_9	.001	.003	.004	.005	.005	.004	.004	.003	.002		k_9 0.15
	k_{10}	.001	.002	.003	.004	.004	.005	.005	.004	.003		k_{10}
0.2	k_9	.002	.004	.006	.006	.006	.006	.005	.004	.002		k_9 0.2
	k_{10}	.001	.003	.004	.005	.006	.006	.006	.005	.003		k_{10}
0.25	k_9	.003	.005	.007	.008	.008	.007	.006	.004			k_9 0.25
	k_{10}	.002	.004	.005	.007	.008	.008	.007	.006			k_{10}
0.3	k_9	.004	.007	.009	.009	.009	.008	.007	.005			k_9 0.3
	k_{10}	.002	.005	.007	.008	.009	.009	.009	.007			k_{10}
0.35	k_9	.005	.008	.010	.011	.011	.009	.007				k_9 0.35
	k_{10}	.003	.006	.008	.010	.011	.011	.010				k_{10}
0.4	k_9	.006	.010	.012	.012	.012	.010	.008				k_9 0.4
	k_{10}	.004	.007	.010	.011	.012	.012	.011				k_{10}
0.45	k_9	.008	.011	.013	.014	.013	.011					k_9 0.45
	k_{10}	.005	.008	.011	.013	.014	.013					k_{10}
0.5	k_9	.009	.013	.015	.015	.014	.012					k_9
	k_{10}	.006	.010	.013	.014	.015	.014					k_{10} 0.5
0.55	k_9	.011	.014	.016	.016	.015						k_9 0.55
	k_{10}	.008	.011	.014	.016	.016						k_{10}
0.6	k_9	.012	.016	.017	.017	.016						k_9 0.6
	k_{10}	.009	.013	.016	.017	.017						k_{10}
0.65	k_9	.014	.017	.019	.018							k_9 0.65
	k_{10}	.010	.014	.017	.018							k_{10}
0.7	k_9	.015	.019	.020	.019							k_9 0.7
	k_{10}	.012	.016	.018	.020							k_{10}
0.75	k_9	.016	.020	.021								k_9 0.75
	k_{10}	.013	.017	.020								k_{10}
0.8	k_9	.018	.021	.022								k_9 0.8
	k_{10}	.014	.018	.021								k_{10}
0.85	k_9	.019	.022									k_9 0.85
	k_{10}	.016	.020									k_{10}
0.9	k_9	.020	.023									k_9 0.9
	k_{10}	.017	.021									k_{10}
0.95	k_9	.021										k_9 0.95
	k_{10}	.018										k_{10}
1.0	k_9	.022										k_9 1.0
	k_{10}	.019										k_{10}

Loading 24 (to Tables 24 A–E)

Free symmetrical triangular line load

$$s = n\,l$$
$$a = m\,l$$

$$0 \leq (m + n) \leq 1.0$$

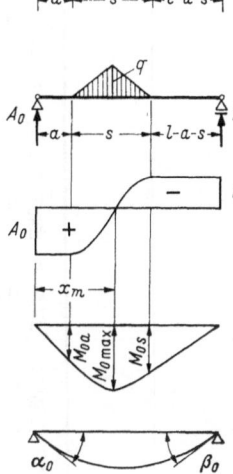

$$M_1 = -k_2\,q\,l^2 \qquad k_1 = \frac{n}{96}\left[48\,m\,(1-m)^2 + n\,(9\,n^2 - 28\,n + 24) + 6\,m\,n\,(12\,m - 16 + 7\,n)\right]$$

$$M_2 = -k_2\,q\,l^2 \qquad k_2 = \frac{n}{96}\left[48\,m^2\,(1-m) + n\,(14\,n - 9\,n^2) + 6\,m\,n\,(8 - 12\,m - 7\,n)\right]$$

$$M_1^0 = -k_1^0\,q\,l^2 \qquad k_1^0 = \frac{n}{64}\left[16\,m\,(m^2 - 3\,m + 2) + n\,(16 - 14\,n + 3\,n^2) + 2\,m\,n\,(12\,m - 24 + 7\,n)\right]$$

$$A_0 = k_3\,q\,l \qquad k_3 = \frac{n}{4}\,(2 - 2\,m - n)$$

$$B_0 = k_4\,q\,l \qquad k_4 = \frac{n}{4}\,(2\,m + n)$$

$$M_{0a} = k_5\,q\,l^2 \qquad k_5 = \frac{m\,n}{4}\,(2 - 2\,m - n)\,; \quad k_5 = m\,k_3$$

$$M_{0s} = k_6\,q\,l^2 \qquad k_6 = \frac{n}{4}\,(2\,m + n)\,(1 - m - n)\,; \quad k_6 = k_4\,(1 - m - n)$$

$$M_{0\,\mathrm{max}} = k_7\,q\,l^2 \qquad k_7 = k_3\,k_8 - \frac{1}{3\,n}\,(k_8^3 - 3\,m\,k_8^2 + 3\,m^2\,k_8 - m^3)$$
$$\text{val. for } \left(m + \frac{n}{2}\right) \leq \frac{1}{2}$$

$$k_7 = k_4\,(1 - k_8) - \frac{1}{3\,n}$$
$$\left[(m + n)^3 - k_8^3 + 3\,(m + n)\,k_8^2 - 3\,(m + n)^2\,k_8\right]$$
$$\text{val. for } \left(m + \frac{n}{2}\right) \leq \frac{1}{2}$$

$$x_m = k_8\,l \qquad k_8 = 0.5\left(2\,m + n\,\sqrt{2 - 2\,m - n}\right)$$
$$\text{val. for } \left(m + \frac{n}{2}\right) \leq \frac{1}{2}$$

$$k_8 = 0.5\left(2\,m + n\left[2 - \sqrt{2\,m + n}\right]\right)$$
$$\text{val. for } \left(m + \frac{n}{2}\right) \leq \frac{1}{2}$$

$$\alpha_0 = k_9\,q\,l^3 \qquad k_9 = \frac{n}{192}\left[16\,m\,(m^2 - 3\,m + 2) + n\,(16 - 14\,n + 3\,n^2) + 2\,m\,n\,(12\,m - 24 + 7\,n)\right]$$

$$\beta_0 = k_{10}\,q\,l^3 \qquad k_{10} = \frac{n}{192}\left[16\,m\,(1 - m^2) + n\,(8 - 3\,n^2) - 2\,m\,n\,(12\,m + 7\,n)\right]$$

For values k_1, k_2 and k_1^0: see Table 24 A

For values k_3 and k_4:　　see Table 24 B

For values k_5 and k_6:　　see Table 24 C

For values k_7 and k_8:　　see Table 24 D

For values k_9 and k_{10}:　　see Table 24 E

Table 24 A   – 68 –

Table 24 A

n		m										n	
		0	0.1	0.2	0.3	0.4	0.5	0.6	0.7	0.8	0.9		
0	k_1	0	0	0	0	0	0	0	0	0	0	k_1	0
	k_2											k_2	
	k_1^0											k_1^0	
0.05	k_1	0.001	0.002	0.003	0.004	0.004	0.002	0.002	0.001	0.001	0.000	k_1	0.05
	k_2	.000	.000	.001	.002	.003	.003	.004	.004	.003	.002	k_2	
	k_1^0	.001	.003	.004	.005	.005	.005	.004	.003	.002	.001	k_1^0	
0.1	k_1	.002	.005	.007	.007	.007	.006	.004	.002	.001	.000	k_1	0.1
	k_2	.000	.001	.002	.004	.006	.007	.007	.007	.005	.002	k_2	
	k_1^0	.002	.004	.008	.009	.010	.009	.008	.006	.004	.001	k_1^0	
0.15	k_1	.005	.009	.011	.011	.010	.008	.005	.003	.001		k_1	0.15
	k_2	.000	.002	.004	.007	.009	.010	.011	.010	.007		k_2	
	k_1^0	.005	.010	.013	.014	.014	.013	.011	.008	.005		k_1^0	
0.2	k_1	.008	.013	.015	.014	.012	.010	.006	.003	.001		k_1	0.2
	k_2	.001	.003	.006	.010	.012	.014	.015	.013	.008		k_2	
	k_1^0	.008	.014	.018	.019	.019	.017	.014	.010	.005		k_1^0	
0.25	k_1	.011	.016	.018	.017	.015	.011	.007	.003			k_1	0.25
	k_2	.002	.005	.009	.013	.017	.018	.018	.014			k_2	
	k_1^0	.012	.019	.023	.024	.023	.020	.016	.011			k_1^0	
0.3	k_1	.015	.020	.022	.020	.017	.012	.007	.003			k_1	0.3
	k_2	.003	.007	.012	.017	.020	.022	.020	.015			k_2	
	k_1^0	.017	.024	.028	.028	.027	.023	.017	.011			k_1^0	
0.35	k_1	.020	.024	.025	.022	.018	.013	.007				k_1	0.35
	k_2	.005	.010	.015	.020	.024	.025	.022				k_2	
	k_1^0	.022	.029	.032	.033	.030	.025	.018				k_1^0	
0.4	k_1	.024	.028	.028	.024	.019	.013	.007				k_1	0.4
	k_2	.007	.013	.019	.024	.028	.028	.024				k_2	
	k_1^0	.027	.034	.032	.037	.033	.027	.019				k_1^0	
0.45	k_1	.028	.031	.030	.026	.020	.013					k_1	0.45
	k_2	.009	.016	.023	.028	.031	.030					k_2	
	k_1^0	.033	.039	.042	.040	.035	.028					k_1^0	
0.5	k_1	.032	.035	.032	.027	.020	.012					k_1	0.5
	k_2	.012	.020	.027	.032	.035	.032					k_2	
	k_1^0	.038	.044	.046	.043	.037	.028					k_1^0	
0.55	k_1	.036	.037	.034	.028	.020						k_1	0.55
	k_2	.016	.024	.031	.036	.037						k_2	
	k_1^0	.044	.049	.050	.046	.038						k_1^0	
0.6	k_1	.039	.040	.035	.028	.019						k_1	0.6
	k_2	.019	.028	.035	.040	.039						k_2	
	k_1^0	.049	.054	.051	.048	.039						k_1^0	
0.65	k_1	.042	.042	.036	.028							k_1	0.65
	k_2	.023	.032	.039	.043							k_2	
	k_1^0	.054	.058	.056	.049							k_1^0	
0.7	k_1	.045	.043	.036	.028							k_1	0.7
	k_2	.028	.036	.043	.045							k_2	
	k_1^0	.059	.061	.058	.050							k_1^0	
0.75	k_1	.047	.044	.037								k_1	0.75
	k_2	.032	.041	.046								k_2	
	k_1^0	.063	.064	.060								k_1^0	
0.8	k_1	.049	.045	.036								k_1	0.8
	k_2	.036	.045	.049								k_2	
	k_1^0	.067	.067	.061								k_1^0	

Table 24 A (Continuation)

n		m										n	
		0	0.1	0.2	0.3	0.4	0.5	0.6	0.7	0.8	0.9		
0.85	k_1	0.050	0.045									k_1	0.85
	k_2	.041	.048									k_2	
	k_1^0	.071	.069									k_1^0	
0.9	k_1	.051	.045									k_1	0.9
	k_2	.045	.051									k_2	
	k_1^0	.074	.070									k_1^0	
0.95	k_1	.052										k_1	0.95
	k_2	.049										k_2	
	k_1^0	.076										k_1^0	
1.0	k_1	.052										k_1	1.0
	k_2	.052										k_2	
	k_1^0	.078										k_1^0	

Table 24B

n		m										n	
		0	0.1	0.2	0.3	0.4	0.5	0.6	0.7	0.8	0.9		
0		0	0	0	0	0	0	0	0	0	0		0
0.05	k_3	0.024	0.022	0.019	0.017	0.014	0.012	0.009	0.007	0.004	0.002	k_3	0.05
	k_4	.001	.003	.006	.008	.011	.013	.016	.018	.021	.023	k_4	
0.1	k_3	.048	.043	.038	.033	.028	.023	.018	.013	.008	.003	k_3	0.1
	k_4	.003	.008	.013	.018	.023	.028	.033	.038	.043	.048	k_4	
0.15	k_3	.069	.062	.054	.047	.039	.032	.024	.017	.009		k_3	0.15
	k_4	.006	.013	.021	.028	.036	.043	.051	.058	.066		k_4	
0.2	k_3	.090	.080	.070	.060	.050	.040	.030	.020	.010		k_3	0.2
	k_4	.010	.020	.030	.040	.050	.060	.070	.080	.090		k_4	
0.25	k_3	.109	.097	.084	.072	.059	.047	.034	.022			k_3	0.25
	k_4	.016	.028	.041	.053	.066	.078	.091	.103			k_4	
0.3	k_3	.128	.113	.098	.083	.068	.053	.038	.023			k_3	0.3
	k_4	.023	.038	.053	.068	.083	.098	.113	.128			k_4	
0.35	k_3	.144	.127	.109	.092	.074	.057	.039				k_3	0.35
	k_4	.031	.048	.066	.083	.101	.118	.136				k_4	
0.4	k_3	.160	.140	.120	.100	.080	.060	.040				k_3	0.4
	k_4	.040	.060	.080	.100	.120	.140	.160				k_4	
0.45	k_3	.174	.152	.129	.107	.084	.062					k_3	0.45
	k_4	.051	.073	.096	.118	.141	.163					k_4	
0.5	k_3	.188	.163	.138	.113	.088	.063					k_3	0.5
	k_4	.063	.088	.113	.138	.163	.188					k_4	
0.55	k_3	.199	.172	.144	.117	.089						k_3	0.55
	k_4	.076	.103	.131	.158	.186						k_4	
0.6	k_3	.210	.180	.150	.120	.090						k_3	0.6
	k_4	.090	.120	.150	.180	.210						k_4	
0.65	k_3	.219	.187	.154	.122							k_3	0.65
	k_1	.106	.138	.171	.203							k_4	
0.7	k_3	.228	.193	.158	.123							k_3	0.7
	k_4	.123	.158	.193	.228								

To be continued Table 24 B on p. 70

Table 24 C   – 70 –

Table 24 B (Continuation)

n		m										n	
		0	0.1	0.2	0.3	0.4	0.5	0.6	0.7*	0.8	0.9		
0.75	k_3	0.234	0.197	0.159								k_3	0.75
	k_4	.141	.178	.216								k_4	
0.8	k_3	.240	.200	.160								k_3	0.8
	k_4	.160	.200	.240								k_4	
0.85	k_3	.244	.202									k_3	0.85
	k_4	.181	.223									k_4	
0.9	k_3	.248	.203									k_3	0.9
	k_4	.202	.248									k_4	
0.95	k_3	.249										k_3	0.95
	k_4	.226										k_4	
1.0	k_3	.250										k_3	1.0
	k_4	.250										k_4	

Table 24 C

n		m										n	
		0	0.1	0.2	0.3	0.4	0.5	0.6	0.7	0.8	0.9		
0	k_5	0	0	0	0	0	0	0	0	0	0	k_5	0
	k_6											k_6	
0.05	k_5	0	0.002	0.004	0.005	0.006	0.006	0.006	0.005	0.004	0.002	k_5	0.05
	k_6	0.001	.003	.004	.005	.006	.006	.005	.005	.003	.001	k_6	
0.1	k_5	0	.004	.008	.010	.011	.011	.011	.009	.006	.002	k_5	0.1
	k_6	.002	.006	.009	.011	.011	.011	.010	.008	.004	0	k_6	
0.15	k_5	0	.006	.011	.014	.016	.016	.015	.012	.008		k_5	0.15
	k_6	.005	.010	.013	.015	.016	.015	.013	.009	.003		k_6	
0.2	k_5	0	.008	.014	.018	.020	.020	.018	.014	.008		k_5	0.2
	k_6	.008	.014	.018	.020	.020	.018	.014	.008	0		k_6	
0.25	k_5	0	.010	.017	.022	.024	.023	.021	.015			k_5	0.25
	k_6	.012	.018	.022	.024	.023	.020	.014	.005			k_6	
0.3	k_5	0	.011	.020	.025	.027	.026	.023	.016			k_5	0.3
	k_6	.016	.023	.026	.027	.025	.020	.011	0			k_6	
0.35	k_5	0	.013	.022	.028	.030	.028	.024				k_5	0.35
	k_6	.020	.026	.030	.029	.025	.018	.007				k_6	
0.4	k_5	0	.014	.024	.030	.032	.030	.024				k_5	0.4
	k_6	.024	.030	.032	.030	.024	.014	0				k_6	
0.45	k_5	0	.015	.026	.032	.034	.031					k_5	0.45
	k_6	.028	.033	.033	.030	.021	.008					k_6	
0.5	k_5	0	.016	.028	.034	.035	.031					k_5	0.5
	k_6	.031	.035	.034	.028	.016	0					k_6	
0.55	k_5	0	.017	.029	.035	.036						k_5	0.55
	k_6	.034	.036	.033	.024	.009						k_6	
0.6	k_5	0	.018	.030	.036	.036						k_5	0.6
	k_6	.036	.036	.030	.018	0						k_6	
0.65	k_5	0	.019	.031	.037							k_5	0.65
	k_6	.037	.035	.026	.010							k_6	

Table 24 C (Continuation)

n		m										n	
		0	0.1	0.2	0.3	0.4	0.5	0.6	0.7	0.8	0.9		
0.7	k_5	0	0.019	0.032	0.037							k_5	0.7
	k_6	.037	.032	.019	0							k_6	
0.75	k_5	0	.020	.032								k_5	0.75
	k_6	.035	.027	.011								k_5	
0.8	k_5	0	.020	.032								k_5	0.8
	k_6	.032	.020	0								k_6	
0.85	k_5	0	.020									k_5	0.85
	k_6	.027	.011									k_6	
0.9	k_5	0	.020									k_5	0.9
	k_6	.020	0									k_6	
0.95	k_5	0										k_5	0.95
	k_6	.011										k_6	
1.0	k_5	0										k_5	1.0
	k_6	0										k_6	

Table 24 D

n		m										n	
		0	0.1	0.2	0.3	0.4	0.5	0.6	0.7	0.8	0.9		
0	k_7	0	0	0	0	0	0	0	0	0	0	k_7	0
	k_8											k_8	
0.05	k_7	0.001	0.003	0.004	0.005	0.006	0.006	0.006	0.005	0.004	0.002	k_7	0.05
	k_8	.044	.138	.233	.330	.427	.524	.622	.719	.815	.910	k_8	
0.1	k_7	.002	.007	.009	.011	.012	.012	.011	.009	.007	.002	k_7	0.1
	k_8	.084	.173	.265	.358	.453	.547	.642	.735	.827	.916	k_8	
0.15	k_7	.005	.010	.014	.017	.024	.017	.016	.012	.008		k_7	0.15
	k_8	.121	.206	.294	.385	.477	.569	.660	.750	.838		k_8	
0.2	k_7	.008	.015	.020	.022	.023	.022	.020	.015	.008		k_7	0.2
	k_8	.155	.237	.323	.411	.500	.589	.677	.763	.845		k_8	
0.25	k_7	.012	.020	.025	.028	.029	.027	.023	.016			k_7	0.25
	k_8	.188	.266	.349	.435	.522	.608	.693	.774			k_8	
0.3	k_7	.017	.025	.031	.033	.033	.031	.025	.017			k_7	0.3
	k_8	.218	.294	.375	.458	.542	.626	.706	.782			k_8	
0.35	k_7	.022	.031	.036	.039	.038	.034	.027				k_7	0.35
	k_8	.246	.320	.398	.479	.561	.641	.717				k_8	
0.4	k_7	.027	.036	.042	.043	.042	.036	.027				k_7	0.4
	k_8	.274	.345	.421	.500	.579	.655	.726				k_8	
0.45	k_7	.033	.042	.047	.048	.045	.038					k_7	0.45
	k_8	.299	.369	.443	.519	.595	.667					k_8	
0.5	k_7	.039	.047	.052	.052	.047	.039					k_7	0.5
	k_8	.323	.391	.463	.537	.609	.677					k_8	
0.55	k_7	.044	.052	.056	.055	.049						k_7	0.55
	k_8	.346	.412	.482	.554	.622						k_8	
0.6	k_7	.050	.057	.060	.057	.050						k_7	0.6
	k_8	.368	.432	.500	.568	.632						k_8	

To be continued Table 24 D on p. 72

Table 24 E – 72 –

Table 24 D (Continuation)

n		m										n
		0	0.1	0.2	0.3	0.4	0.5	0.6	0.7	0.8	0.9	
0.65	k_7	0.055	0.062	0.063	0.059							k_7 0.65
	k_8	.388	.450	.517	.581							k_8
0.7	k_7	.061	.066	.066	.061							k_7 0.7
	k_8	.407	.468	.532	.593							k_8
0.75	k_7	.066	.070	.069								k_7 0.75
	k_8	.425	.485	.546								k_8
0.8	k_7	.070	.073	.070								k_7 0.8
	k_8	.442	.500	.558								k_8
0.85	k_7	.074	.076									k_7 0.85
	k_8	.458	.514									k_8
0.9	k_7	.078	.078									k_7 0.9
	k_8	.473	.527									k_8
0.95	k_7	.081										k_7 0.95
	k_8	.487										k_8
1.0	k_7	.083										k_7 1.0
	k_8	.500										k_8

Table 24E

n		m										n
		0	0.1	0.2	0.3	0.4	0.5	0.6	0.7	0.8	0.9	
0	k_9	0	0	0	0	0	0	0	0	0	0	k_9 0
	k_{10}	0	0	0	0	0	0	0	0	0	0	k_{10}
0.05	k_9	0.000	0.001	0.001	0.002	0.002	0.002	0.001	0.001	0.001	0.000	k_9 0.05
	k_{10}	.000	.001	.001	.001	.001	.002	.002	.001	.001	.001	k_{10}
0.1	k_9	.001	.002	.003	.003	.003	.003	.003	.002	.001	.000	k_9 0.1
	k_{10}	.000	.001	.002	.003	.003	.003	.003	.003	.002	.001	k_{10}
0.15	k_9	.002	.003	.004	.005	.005	.004	.004	.003	.002		k_9 0.15
	k_{10}	.001	.002	.003	.004	.005	.005	.005	.004	.003		k_{10}
0.2	k_9	.003	.005	.006	.006	.006	.006	.005	.003	.002		k_9 0.2
	k_{10}	.002	.003	.005	.006	.006	.006	.006	.005	.003		k_{10}
0.25	k_9	.004	.006	.008	.008	.008	.007	.005	.004			k_9 0.25
	k_{10}	.003	.004	.006	.007	.008	.008	.007	.005			k_{10}
0.3	k_9	.006	.008	.009	.009	.009	.008	.006	.004			k_9 0.3
	k_{10}	.004	.006	.008	.009	.009	.009	.008	.006			k_{10}
0.35	k_9	.007	.010	.015	.011	.010	.008	.006				k_9 0.35
	k_{10}	.005	.007	.009	.011	.011	.010	.009				k_{10}
0.4	k_9	.009	.011	.012	.012	.011	.009	.006				k_9 0.4
	k_{10}	.006	.009	.011	.012	.012	.011	.009				k_{10}
0.45	k_9	.011	.013	.014	.013	.012	.009					k_9 0.45
	k_{10}	.008	.011	.013	.014	.014	.012					k_{10}
0.5	k_9	.013	.015	.015	.015	.012	.009					k_9 0.5
	k_{10}	.009	.012	.015	.015	.015	.013					k_{10}
0.55	k_9	.015	.016	.017	.015	.013						k_9 0.55
	k_{10}	.011	.014	.016	.017	.016						k_{10}

Table 24 E (Continuation)

n		m											n	
		0	0.1	0.2	0.3	0.4	0.5	0.6	0.7	0.8	0.9			
0.6	k_9	0.016	0.018	0.018	0.016	0.013							k_9	0.6
	k_{10}	.013	.016	.018	.018	.016							k_{10}	
0.65	k_9	0.18	.019	.019	.016								k_9	0.65
	k_{10}	.015	.018	.019	.019								k_{10}	
0.7	k_9	.020	.020	.019	.017								k_9	0.7
	k_{10}	.017	.019	.020	.020								k_{10}	
0.75	k_9	.021	.021	.020									k_9	0.75
	k_{10}	.018	.021	.022									k_{10}	
0.8	k_9	.022	.022	.020									k_9	0.8
	k_{10}	.020	.022	.022									k_{10}	
0.85	k_9	.024	.023										k_9	0.85
	k_{10}	.022	.024										k_{10}	
0.9	k_9	.025	.024										k_9	0.9
	k_{10}	.024	.025										k_{10}	
0.95	k_9	.025											k_9	0.95
	k_{10}	.025											k_{10}	
1.0	k_9	.026											k_9	1.0
	k_{10}	.026											k_{10}	

Table 25 A   – 74 –

Loading 25 (to Tables 25 A–B)

Free, symmetrical twin triangular line load

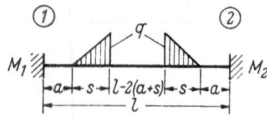

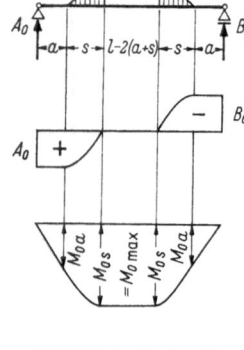

$$s = n\,l \qquad\qquad 0 \le n \le 0.5$$

$$a = m\,l \qquad\qquad 0 \le m \le 0.4$$

$$M_1 = -k_1 q\,l \qquad k_1 = \frac{n}{12}\left[2\,m\,(3 - 3\,m - 4\,n) + n\,(4 - 3\,n)\right]$$

$$M_2 = M_1$$

$$M_1^0 = -k_1^0 q\,l^2 \qquad k_1^0 = \frac{n}{8}\left[2\,m\,(3 - 3\,m - 4\,n) + n\,(4 - 3\,n)\right]$$

$$k_1^0 = 1.5\,k_1$$

$$A_0 = k_3 q\,l \qquad k_3 = \frac{n}{2}$$

$$B_0 = A_0$$

$$M_{0\,a} = k_5 q\,l^2 \qquad k_5 = \frac{m\,n}{2}$$

$$M_{0\,s} = k_6 q\,l^2 \qquad k_6 = \frac{n}{6}\,(3\,m + 2\,n)$$

$$M_{0\,\max} = M_{0\,s}$$

$$(a + s) \le x_m \le l - (a + s)$$

$$\alpha_0 = k_9 q\,l^3 \qquad k_9 = \frac{n}{24}\left[2\,m\,(3 - 3\,m - 4\,n) + n\,(4 - 3\,n)\right]$$

$$\beta_0 = \alpha_0 \qquad k_9 = \frac{1}{2}\,k_1$$

For values k_1, k_1^0 and k_3: see Table 25 A

For values k_5, k_6 and k_9: see Table 25 B

Table 25 A

n		m											n	
		0	0.05	0.1	0.15	0.2	0.25	0.3	0.35	0.4	0.45	0.5		
0	k_1	0	0	0	0	0	0	0	0	0	0	0	k_1	0
	k_1^0												k_1^0	
0.05	k_1	0.001	0.002	0.003	0.004	0.004	0.005	0.006	0.006	0.006	0.006		k_1	0.05
	k_1^0	.001	.003	.004	.006	.007	.008	.008	.009	.009	.009		k_1^0	
0.1	k_1	.003	.005	.007	.008	.010	.011	.012	.012	.012			k_1	0.1
	k_1^0	.005	.008	.010	.013	.015	.016	.017	.018	.019			k_1^0	
0.15	k_1	.007	.009	.012	.014	.016	.017	.018	.018				k_1	0.15
	k_1^0	.010	.014	.018	.021	.023	.025	.027	.028				k_1^0	
0.2	k_1	.011	.015	.018	.020	.022	.023	.024					k_1	0.2
	k_1^0	.017	.022	.027	.030	.033	.035	.037					k_1^0	
0.25	k_1	.017	.021	.024	.027	.029	.030						k_1	0.25
	k_1^0	.025	.031	.036	.040	.043	.045						k_1^0	
0.3	k_1	.023	.027	.031	.033	.035							k_1	0.3
	k_1^0	.035	.041	.046	.049	.053							k_1^0	
0.35	k_1	.030	.034	.038	.040								k_1	0.35
	k_1^0	.045	.052	.057	.060								k_1^0	

Table 25 A (Continuation)

n		m												n
		0	0.05	0.1	0.15	0.2	0.25	0.3	0.35	0.4	0.45	0.5		
0.4	k_1	0.037	0.041	0.045									k_1	0.4
	k_1^0	.056	.062	.067									k_1^0	
0.45	k_1	.045	.049										k_1	0.45
	k_1^0	.067	.073										k_1^0	
0.5	k_1	.052											k_1	0.5
	k_1^0	.078											k_1^0	

n	0	0.05	0.1	0.15	0.2	0.25	0.3	0.35	0.4	0.45	0.5	n
k_3	0	0.025	0.050	0.075	0.100	0.125	0.150	0.175	0.200	0.225	0.250	k_3

Table 25 B

n		m												n
		0	0.05	0.1	0.15	0.2	0.25	0.3	0.35	0.4	0.45	0.5		
0	k_5	0	0	0	0	0	0	0	0	0	0	0	k_5	0
	k_6												k_6	
	k_9												k_9	
0.05	k_5	0	0.001	0.003	0.004	0.005	0.006	0.008	0.009	0.010	0.011		k_5	0.05
	k_6	0.001	.002	.003	.005	.006	.007	.008	.010	.011	.012		k_6	
	k_9	.000	.001	.001	.002	.002	.003	.003	.003	.003	.003		k_9	
0.1	k_5	0	.003	.005	.008	.010	.013	.015	.018	.020			k_5	0.1
	k_6	.003	.006	.008	.011	.013	.016	.018	.021	.023			k_6	
	k_9	.002	.003	.003	.004	.005	.005	.006	.006	.006			k_9	
0.15	k_5	0	.004	.008	.011	.015	.019	.023	.026				k_5	0.15
	k_6	.008	.011	.015	.019	.023	.026	.030	.034				k_6	
	k_9	.003	.005	.006	.007	.008	.008	.009	.009				k_9	
0.2	k_5	0	.005	.010	.015	.020	.025	.030					k_5	0.2
	k_6	.013	.018	.023	.028	.033	.038	.043					k_6	
	k_9	.006	.007	.009	.010	.011	.012	.012					k_9	
0.25	k_5	0	.006	.013	.019	.025	.031						k_5	0.25
	k_6	.021	.027	.033	.040	.046	.052						k_6	
	k_9	.008	.010	.012	.013	.014	.015						k_9	
0.3	k_5	0	.008	.015	.023	.030							k_5	0.3
	k_6	.030	.038	.045	.053	.060							k_6	
	k_9	.012	.014	.015	.016	.018							k_9	
0.35	k_5	0	.009	.018	.026								k_5	0.35
	k_6	.041	.050	.058	.067								k_6	
	k_9	.015	.017	.019	.020								k_9	
0.4	k_5	0	.010	.020									k_5	0.4
	k_6	.053	.063	.073									k_6	
	k_9	.019	.021	.022									k_9	
0.45	k_5	0	.011										k_5	0.45
	k_6	.068	.079										k_6	
	k_9	.022	.024										k_9	
0.5	k_5	0											k_5	0.5
	k_6	.083											k_6	
	k_9	.026											k_9	

Loading 26 (to Tables 26 A–E)

Free line load according to a quadratic parabola

$$a = m\,l \qquad\qquad s = n\,l$$

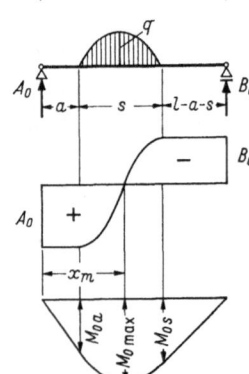

$$M_1 = -k_1\,q\,l^2 \qquad k_1 = \frac{1}{15}\left\{30\,k_5\,(m - m^2) + 15\,k_6 \times\right.$$
$$\times\,[1 - 3\,n + 2\,n^2 - m\,(3 - 2\,m - 4\,n)] +$$
$$+\,30\,n\,k_3\,[n - n^2 + m\,(2 - 3\,m - 3\,n)] -$$
$$\left. -\,n^3\,(6 - 9\,m - 7\,n)\right\}$$

$$M_2 = -k_2\,q\,l^2 \qquad k_2 = \frac{1}{15}\left\{15\,k_5\,(2\,m^2 - m) + 30\,k_6 \times\right.$$
$$\times\,[n - n^2 + m\,(1 - m - 2\,n)] +$$
$$+\,15\,n\,k_3\,[2\,n^2 - n - 2\,m\,(1 - 3\,m - 3\,n) +$$
$$\left. +\,n^3\,(3 - 9\,m - 7\,n)]\right\}$$

$$M_1^0 = -k_1^0\,q\,l^2 \qquad k_1^0 = 3\,k_9$$

$$A_0 = k_3\,q\,l \qquad k_3 = \frac{n}{3}\,(2 - 2\,m - n)$$

$$B_0 = k_4\,q\,l \qquad k_4 = \frac{n}{3}\,(n + 2\,m)$$

$$M_{0\,a} = k_5\,q\,l^2 \qquad k_5 = \frac{m\,n}{3}\,(2 - 2\,m - n) \qquad k_5 = m\,k_3$$

$$M_{0\,s} = k_6\,q\,l^2 \qquad k_6 = \frac{n}{3}\,[n - n^2 + m\,(2 - 2\,m - 3\,n)]$$
$$k_6 = k_4\,(1 - m - n)$$

$$M_{0\,\max} = k_7\,q\,l^2 \qquad k_7 = k_3\,k_8 - \frac{(k_8 - m)^3}{3\,n^2}\,(2\,n + m - k_8)$$

$$x_m = k_8\,l \qquad k_8 = m + \frac{n}{2} \times$$
$$\times\left\{1 - 2\cos\left[\frac{1}{3}\arccos\,(2\,m + n - 1) + 60°\right]\right\}$$

$$\alpha_0 = k\,q_9\,l^3 \qquad k_9 = \frac{1}{90}\left\{15\,k_5\,(3\,m - 2\,m^2) + 30\,k_6\,(1 - m - n)^2 +\right.$$
$$+\,15\,k_3\,[6\,n\,(m - m^2) + n^2\,(3 - 6\,m - 2\,n)] -$$
$$\left. -\,9\,n^3\,(1 - m) + 7\,n^4\right\}$$

$$\beta_0 = k_{10}\,q\,l^3 \qquad k_{10} = \frac{1}{90}\left\{30\,k_5\,m^2 + 15\,k_6\,[1 + n - 2\,n^2 +\right.$$
$$+\,m\,(1 - 2\,m - 4\,n] + 30\,k_3\,(n^3 + 3\,m\,n^2 +$$
$$\left. +\,3\,m^2\,n) - 9\,m\,n^3 - 7\,n^4\right\}$$

For values k_1, k_2 and k_1^0: see Table 26 A
For values k_3 and k_4:   see Table 26 B
For values k_5 and k_6:   see Table 26 C
For values k_7 and k_8:   see Table 26 D
For values k_9 and k_{10}:   see Table 26 E

Table 26A

n		m											n
		0	0.1	0.2	0.3	0.4	0.5	0.6	0.7	0.8	0.9	1.0	
0	k_1	0	0	0	0	0	0	0	0	0	0	0	k_1
	k_2												k_2 0
	k_1^0												k_1^0
0.05	k_1	0.001	0.003	0.005	0.005	0.005	0.004	0.003	0.002	0.001	0.000		k_1
	k_2	.000	.000	.001	.002	.003	.004	.005	.005	.004	.002		k_2 0.05
	k_1^0	.001	.003	.005	.006	.006	.006	.005	.004	.003	.001		k_1^0
0.1	k_1	.003	.007	.009	.010	.009	.007	.005	.003	.001	.000		k_1
	k_2	.000	.001	.003	.005	.007	.009	.010	.009	.007	.003		k_2 0.1
	k_1^0	.003	.008	.011	.012	.013	.012	.010	.008	.005	.002		k_1^0
0.15	k_1	.006	.012	.014	.015	.013	.010	.007	.004	.001			k_1
	k_2	.001	.003	.006	.009	.012	.014	.015	.013	.009			k_2 0.15
	k_1^0	.007	.013	.017	.019	.019	.017	.014	.011	.006			k_1^0
0.2	k_1	.010	.017	.019	.019	.017	.013	.008	.004	.001			k_1
	k_2	.001	.004	.008	.013	.017	.019	.019	.017	.010			k_2 0.2
	k_1^0	.011	.019	.024	.025	.025	.022	.018	.013	.007			k_1^0
0.25	k_1	.015	.022	.024	.023	.020	.015	.009	.004				k_1
	k_2	.003	.007	.012	.017	.022	.024	.023	.019				k_2 0.25
	k_1^0	.016	.025	.030	.032	.030	.027	.021	.014				k_1^0
0.3	k_1	.020	.027	.029	.027	.022	.016	.010	.004				k_1
	k_2	.004	.010	.016	.022	.027	.029	.027	.020				k_2 0.3
	k_1^0	.022	.032	.037	.038	.035	.030	.023	.014				k_1^0
0.35	k_1	.026	.032	.033	.030	.024	.017	.010					k_1
	k_2	.007	.013	.020	.027	.032	.033	.030					k_2 0.35
	k_1^0	.029	.039	.043	.043	.040	.033	.024					k_1^0
0.4	k_1	.031	.037	.037	.032	.025	.017	.009					k_1
	k_2	.009	.017	.025	.032	.037	.037	.031					k_2 0.4
	k_1^0	.036	.045	.049	.048	.044	.035	.025					k_1^0
0.45	k_1	.037	.041	.040	.034	.026	.017						k_1
	k_2	.013	.021	.030	.038	.041	.040						k_2 0.45
	k_1^0	.043	.052	.055	.053	.047	.037						k_1^0
0.5	k_1	.042	.045	.043	.036	.026	.017						k_1
	k_2	.017	.026	.036	.043	.045	.042						k_2 0.5
	k_1^0	.050	.059	.061	.057	.049	.038						k_1^0
0.55	k_1	.046	.049	.045	.037	.026							k_1
	k_2	.021	.032	.041	.048	.049							k_2 0.55
	k_1^0	.057	.065	.065	.060	.051							k_1^0
0.6	k_1	.051	.052	.046	.037	.026							k_1
	k_2	.026	.037	.046	.052	.051							k_2 0.6
	k_1^0	.064	.070	.070	.063	.051							k_1^0
0.65	k_1	.055	.054	.047	.037								k_1
	k_2	.031	.042	.051	.055								k_2 0.65
	k_1^0	.070	.076	.073	.065								k_1^0
0.7	k_1	.058	.056	.048	.037								k_1
	k_2	.037	.048	.056	.058								k_2 0.7
	k_1^0	.076	.080	.076	.066								k_1^0
0.75	k_1	.061	.057	.048									k_1
	k_2	.042	.053	.060									k_2 0.75
	k_1^0	.082	.084	.078									k_1^0
0.8	k_1	.063	.058	.048									k_1
	k_2	.048	.058	.063									k_2 0.8
	k_1^0	.087	.087	.079									k_1^0

6*

To be continued Table 26 A on p. 78

Table 26 B   – 78 –

Table 26 A (Continuation)

n		m											n	
		0	0.1	0.2	0.3	0.4	0.5	0.6	0.7	0.8	0.9	1.0		
0.85	k_1 k_2 k_1^0	0.065 .053 .091	0.058 .062 .090										k_1 k_2 k_1^0	0.85
0.9	k_1 k_2 k_1^0	.066 .058 .095	.058 .066 .091										k_1 k_2 k_1^0	0.9
0.95	k_1 k_2 k_1^0	.066 .063 .098											k_1 k_2 k_1^0	0.95
1.0	k_1 k_2 k_1^0	.067 .067 .100											k_1 k_2 k_1^0	1.0

Table 26 B

n		m											n	
		0	0.1	0.2	0.3	0.4	0.5	0.6	0.7	0.8	0.9	1.0		
0	k_3 k_4	0 0	0 0	0 0	0 0	0 0	0 0	0 0	0 0	0 0	0 0	0 0	k_3 k_4	0
0.05	k_3 k_4	0.033 .001	0.029 .004	0.026 .008	0.023 .011	0.019 .014	0.016 .018	0.013 .021	0.009 .024	0.006 .028	0.003 .031		k_3 k_4	0.05
0.1	k_3 k_4	.063 .003	.057 .010	.050 .017	.043 .023	.037 .030	.030 .037	.023 .043	.017 .050	.010 .057	.003 .063		k_3 k_4	0.1
0.15	k_3 k_4	.093 .008	.083 .018	.073 .028	.063 .038	.053 .048	.043 .058	.033 .068	.023 .078	.013 .088			k_3 k_4	0.15
0.2	k_3 k_4	.120 .013	.107 .027	.093 .040	.080 .053	.067 .067	.053 .080	.040 .093	.027 .107	.013 .120			k_3 k_4	0.2
0.25	k_3 k_4	.146 .021	.129 .038	.113 .054	.096 .071	.079 .088	.063 .104	.046 .121	.029 .138				k_3 k_4	0.25
0.3	k_3 k_4	.170 .030	.150 .050	.130 .070	.110 .090	.090 .110	.070 .130	.050 .150	.030 .170				k_3 k_4	0.3
0.35	k_3 k_4	.193 .041	.169 .064	.146 .088	.122 .111	.099 .134	.076 .157	.053 .181					k_3 k_4	0.35
0.4	k_3 k_4	.213 .053	.187 .080	.160 .107	.133 .133	.107 .160	.080 .187	.053 .213					k_3 k_4	0.4
0.45	k_3 k_4	.233 .068	.203 .098	.173 .128	.143 .158	.113 .188	.083 .218						k_3 k_4	0.45
0.5	k_3 k_4	.250 .083	.217 .117	.183 .150	.150 .183	.117 .217	.083 .250						k_3 k_4	0.5
0.55	k_3 k_4	.266 .101	.229 .138	.193 .174	.156 .211	.119 .248							k_3 k_4	0.55
0.6	k_3 k_4	.280 .120	.240 .160	.200 .200	.160 .240	.120 .280							k_3 k_4	0.6

Table 26 B (Continuation)

n		m 0	0.1	0.2	0.3	0.4	0.5	0.6	0.7	0.8	0.9	1.0		n
0.65	k_3	.292	.249	.206	.163								k_3	0.65
	k_4	.141	.184	.227	.271								k_4	
0.7	k_3	.303	.257	.210	.163								k_3	0.7
	k_4	.163	.210	.257	.303								k_4	
0.75	k_3	.313	.263	.213									k_3	0.75
	k_4	.188	.238	.288									k_4	
0.8	k_3	.320	.267	.213									k_3	0.8
	k_4	.213	.267	.320									k_4	
0.85	k_3	.326	.269										k_3	0.85
	k_4	.241	.298										k_4	
0.9	k_3	.330	.270										k_3	0.9
	k_4	.270	.330										k_4	
0.95	k_3	.332											k_3	0.95
	k_4	.301											k_4	
1.0	k_3	.333											k_3	1.0
	k_4	.333											k_4	

Table 26 C

n		m 0	0.1	0.2	0.3	0.4	0.5	0.6	0.7	0.8	0.9	1.0		n
0	k_5	0	0	0	0	0	0	0	0	0	0	0	k_5	0
	k_6												k_6	
0.05	k_5	0	0.003	0.005	0.007	0.008	0.008	0.008	0.006	0.005	0.002		k_5	0.05
	k_6	0.001	.004	.006	.007	.008	.008	.007	.006	.004	.002		k_6	
0.1	k_5	0	.006	.010	.013	.015	.015	.014	.012	.008	.003		k_5	0.1
	k_6	0.003	.008	.012	.014	.015	.015	.013	.010	.006	0		k_6	
0.15	k_5	0	.008	.015	.019	.021	.021	.020	.016	.010			k_5	0.15
	k_6	0.006	.013	.018	.021	.021	.020	.017	.012	.004			k_6	
0.2	k_5	0	.011	.019	.024	.027	.027	.024	.019	.011			k_5	0.2
	k_6	0.011	.019	.024	.027	.027	.024	.019	.011	0			k_6	
0.25	k_5	0	.013	.023	.029	.032	.031	.028	.020				k_5	0.25
	k_6	0.016	.024	.030	.032	.031	.026	.018	.007				k_6	
0.3	k_5	0	.015	.026	.033	.036	.035	.030	.021				k_5	0.3
	k_6	0.021	.030	.035	.036	.033	.026	.015	0				k_6	
0.35	k_5	0	.017	.029	.037	.040	.038	.032					k_5	0.35
	k_6	0.027	.035	.039	.039	.034	.024	.009					k_6	
0.4	k_5	0	.019	.032	.040	.043	.040	.032					k_5	0.4
	k_6	0.032	.040	.043	.040	.032	.019	0					k_6	
0.45	k_5	0	.020	.035	.043	.045	.041						k_5	0.45
	k_6	0.037	.044	.045	.039	.028	.011						k_6	
0.5	k_5	0	.022	.037	.045	.047	.042						k_5	0.5
	k_6	0.042	.047	.045	.037	.022	0						k_6	

To be continued Table 26 C on p. 80

Table 26 D   – 80 –

Table 26 C (Continuation)

n		m											n	
		0	0,1	0.2	0.3	0.4	0.5	0.6	0.7	0.8	0.9	1.0		
0.55	k_5	0	.023	.039	.047	.048							k_5	0.55
	k_6	0.045	.048	.044	.032	.012							k_6	
0.6	k_5	0	.024	.040	.048	.048							k_5	0.6
	k_6	0.048	.048	.040	.024	0							k_6	
0.65	k_5	0	.025	.041	.049								k_5	0.65
	k_6	0.049	.046	.034	.014								k_6	
0.7	k_5	0	.026	.042	.049								k_5	0.7
	k_6	0.049	.042	.026	0								k_6	
0.75	k_5	0	.026	.043									k_5	0.75
	k_6	0.047	.036	.014									k_6	
0.8	k_5	0	.027	.043									k_5	0.8
	k_6	0.043	.027	0									k_6	
0.85	k_5	0	.027										k_5	0.85
	k_6	0.036	.015										k_6	
0.9	k_5	0	.027										k_5	0.9
	k_6	0.027	0										k_6	
0.95	k_5	0											k_5	0.95
	k_6	0.015											k_6	
1.0	k_5	0											k_5	1.0
	k_6	0											k_6	

Table 26 D

n		m											n	
		0	0.1	0.2	0.3	0.4	0.5	0.6	0.7	0.8	0.9	1.0		
0	k_7	0	0	0	0	0	0	0	0	0	0	0	k_7	0
	k_8	–	–	–	–	–	–	–	–	–	–	–	k_8	
0.05	k_7	0.001	0.004	0.006	0.007	0.008	0.008	0.008	0.007	0.005	0.002		k_7	0.05
	k_8	.045	.139	.235	.331	.428	.524	.621	.717	.813	.908		k_8	
0.1	k_7	.003	.008	.012	.015	.016	.016	.015	.012	.008	.003		k_7	0.1
	k_8	.086	.176	.267	.360	.453	.547	.640	.733	.824	.914		k_8	
0.15	k_7	.006	.014	.019	.022	.024	.023	.021	.016	.010			k_7	0.15
	k_8	.125	.208	.298	.388	.478	.567	.657	.746	.833			k_8	
0.2	k_7	.011	.020	.026	.030	.031	.030	.026	.020	.011			k_7	0.2
	k_8	.161	.243	.327	.413	.500	.587	.673	.757	.839			k_8	
0.25	k_7	.016	.026	.033	.037	.038	.035	.030	.022				k_7	0.25
	k_8	.195	.273	.355	.438	.521	.604	.686	.767				k_8	
0.3	k_7	.022	.033	.040	.044	.044	.040	.033	.022				k_7	0.3
	k_8	.227	.302	.380	.460	.540	.620	.698	.773				k_8	
0.35	k_7	.029	.040	.047	.051	.050	.044	.035					k_7	0.35
	k_8	.257	.329	.404	.481	.557	.633	.707					k_8	
0.4	k_7	.036	.047	.054	.057	.054	.047	.036					k_7	0.4
	k_8	.285	.355	.427	.500	.573	.645	.715					k_8	

Table 26 D (Continuation)

n		m											n	
		0	0.1	0.2	0.3	0.4	0.5	0.6	0.7	0.8	0.9	1.0		
0.45	k_7	0.043	0.055	0.061	0.062	0.058	0.049						k_7	0.45
	k_8	.312	.379	.448	.517	.587	.655						k_8	
0.5	k_7	.050	.061	.067	.067	.061	.050						k_7	0.5
	k_8	.337	.401	.467	.533	.599	.663						k_8	
0.55	k_7	.058	.068	.073	.071	.064							k_7	0.55
	k_8	.360	.421	.484	.547	.610							k_8	
0.6	k_7	.065	.074	.078	.074	.065							k_7	0.6
	k_8	.382	.440	.500	.560	.618							k_8	
0.65	k_7	.072	.080	.082	.077								k_7	0.65
	k_8	.402	.458	.514	.570								k_8	
0.7	k_7	.078	.085	.085	.078								k_7	0.7
	k_8	.421	.473	.527	.579								k_8	
0.75	k_7	.084	.090	.088									k_7	0.75
	k_8	.438	.487	.537									k_8	
0.8	k_7	.086	.093	.089									k_7	0.8
	k_8	.454	.500	.546									k_8	
0.85	k_7	.094	.096										k_7	0.85
	k_8	.468	.511										k_8	
0.9	k_7	.098	.098										k_7	0.9
	k_8	.480	.520										k_8	
0.95	k_7	.102											k_7	0.95
	k_8	.491											k_8	
1.0	k_7	.104											k_7	1.0
	k_8	.500											k_8	

Table 26 E

n		m											n	
		0	0.1	0.2	0.3	0.4	0.5	0.6	0.7	0.8	0.9	1.0		
0	k_9	0	0	0	0	0	0	0	0	0	0	0	k_9	0
	k_{10}												k_{10}	
0.05	k_9	0.000	0.001	0.002	0.002	0.002	0.002	0.002	0.001	0.001	0.000		k_9	0.05
	k_{10}	.000	.001	.001	.002	.002	.002	.002	.002	.001	.001		k_{10}	
0.1	k_9	.001	.003	.004	.004	.004	.004	.003	.003	.002	.001		k_9	0.1
	k_{10}	.001	.002	.003	.003	.004	.004	.004	.004	.003	.001		k_{10}	
0.15	k_9	.002	.004	.006	.006	.006	.006	.005	.004	.002			k_9	0.15
	k_{10}	.001	.003	.004	.005	.006	.006	.006	.005	.003			k_{10}	
0.2	k_9	.004	.006	.008	.008	.008	.007	.006	.004	.002			k_9	0.2
	k_{10}	.002	.004	.006	.007	.008	.008	.008	.006	.004			k_{10}	
0.25	k_9	.005	.008	.010	.011	.010	.009	.007	.005				k_9	0.25
	k_{10}	.003	.006	.008	.010	.010	.010	.009	.007				k_{10}	
0.3	k_9	.007	.011	.012	.013	.012	.010	.008	.005				k_9	0.3
	k_{10}	.005	.008	.010	.012	.013	.012	.011	.007				k_{10}	

To be continued Table 26 E on p. 82

Table 26 E   – 82 –

Table 26E (Continuation)

n		m 0	0.1	0.2	0.3	0.4	0.5	0.6	0.7	0.8	0.9	1.0		n
0.35	k_9	0.010	0.013	0.014	0.014	0.013	0.011	0.008					k_9	0.35
	k_{10}	.006	.010	.012	.014	.015	.014	.011					k_{10}	
0.4	k_9	.012	.015	.016	.016	.015	.012	.008					k_9	0.4
	k_{10}	.008	.012	.015	.016	.016	.015	.012					k_{10}	
0.45	k_9	.014	.017	.018	.018	.016	.012						k_9	0.45
	k_{10}	.010	.014	.017	.018	.018	.016						k_{10}	
0.5	k_9	.017	.020	.020	.019	.016	.013						k_9	0.5
	k_{10}	.013	.016	.019	.020	.020	.017						k_{10}	
0.55	k_9	.019	.022	.022	.020	.017							k_9	0.55
	k_{10}	.015	.019	.021	.022	.021							k_{10}	
0.6	k_9	.021	.023	.023	.021	.017							k_9	0.6
	k_{10}	.017	.021	.023	.023	.021							k_{10}	
0.65	k_9	.023	.025	.024	.022								k_9	0.65
	k_{10}	.020	.023	.025	.025								k_{10}	
0.7	k_9	.025	.027	.025	.022								k_9	0.7
	k_{10}	.022	.025	.027	.025								k_{10}	
0.75	k_9	.027	.028	.026									k_9	0.75
	k_{10}	.024	.027	.028									k_{10}	
0.8	k_9	.029	.029	.026									k_9	0.8
	k_{10}	.026	.029	.029									k_{10}	
0.85	k_9	.030	.030										k_9	0.85
	k_{10}	.029	.031										k_{10}	
0.9	k_9	.032	.030										k_9	0.9
	k_{10}	.030	.032										k_{10}	
0.95	k_9	.033											k_9	0.95
	k_{10}	.032											k_{10}	
1.0	k_9	.033											k_9	1.0
	k_{10}	.033											k_{10}	

Loading 27 (to Table 27)

Concentrated load in any position

$$a = m \, l$$

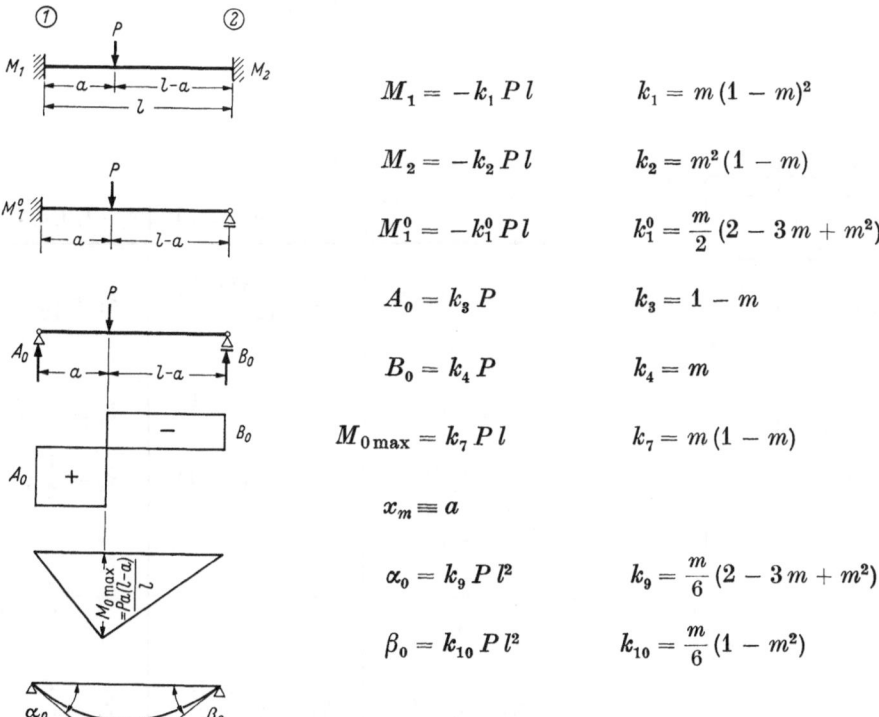

$$M_1 = - k_1 \, P \, l \qquad k_1 = m \, (1 - m)^2$$

$$M_2 = - k_2 \, P \, l \qquad k_2 = m^2 \, (1 - m)$$

$$M_1^0 = - k_1^0 \, P \, l \qquad k_1^0 = \frac{m}{2} \, (2 - 3 \, m + m^2)$$

$$A_0 = k_3 \, P \qquad k_3 = 1 - m$$

$$B_0 = k_4 \, P \qquad k_4 = m$$

$$M_{0\,\text{max}} = k_7 \, P \, l \qquad k_7 = m \, (1 - m)$$

$$x_m \equiv a$$

$$\alpha_0 = k_9 \, P \, l^2 \qquad k_9 = \frac{m}{6} \, (2 - 3 \, m + m^2)$$

$$\beta_0 = k_{10} \, P \, l^2 \qquad k_{10} = \frac{m}{6} \, (1 - m^2)$$

Table 27

m	k_1	k_2	k_1^0	k_3	k_4	k_7	k_9	k_{10}	m
0	0	0	0	1.00	0	0	0	0	0
0.05	0.045	0.002	0.046	0.95	0.05	0.048	0.015	0.008	0.05
0.1	.081	.009	.086	.90	.10	.090	.029	.017	0.1
0.15	.108	.019	.118	.85	.15	.128	.039	.024	0.15
0.2	.128	.032	.144	.80	.20	.160	.048	.032	0.2
0.25	.141	.047	.164	.75	.25	.188	.055	.039	0.25
0.3	.147	.063	.179	.70	.30	.210	.060	.046	0.3
0.35	.148	.080	.188	.65	.35	.228	.063	.051	0.35
0.4	.144	.096	.192	.60	.40	.240	.064	.056	0.4
0.45	.136	.111	.192	.55	.45	.248	.064	.042	0.45
0.5	.125	.125	.188	.50	.50	.250	.063	.063	0.5
0.55	.111	.136	.179	.45	.55	.248	.060	.064	0.55
0.6	.096	.144	.168	.40	.60	.240	.056	.064	0.6
0.65	.080	.148	.154	.35	.65	.228	.051	.063	0.65
0.7	.063	.147	.137	.30	.70	.210	.046	.060	0.7
0.75	.047	.141	.117	.25	.75	.188	.039	.055	0.75
0.8	.032	.128	.096	.20	.80	.160	.032	.048	0.8
0.85	.019	.108	.073	.15	.85	.128	.024	.039	0.85
0.9	.009	.081	.050	.10	.90	.090	.017	.029	0.9
0.95	.002	.045	.025	.05	.95	.048	.008	.015	0.95
1.0	0.000	0.000	0.000	0.000	1.00	0.000	0.000	0.000	1.00

Table 28   – 84 –

Loading 28 (to Table 28)

Several symmetrically arranged concentrated loads, dividing the span into equal intervals

$$a = \frac{l}{r}$$

$(r - 1)$ loads P being present.

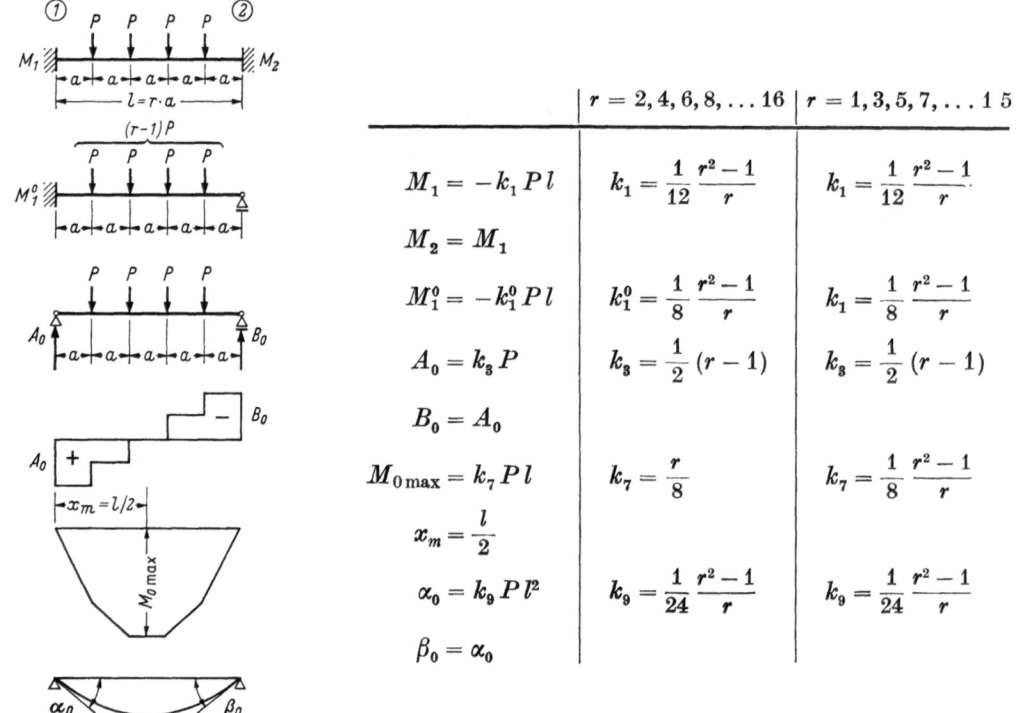

	$r = 2, 4, 6, 8, \ldots 16$	$r = 1, 3, 5, 7, \ldots 15$
$M_1 = -k_1 \, P \, l$	$k_1 = \dfrac{1}{12} \dfrac{r^2 - 1}{r}$	$k_1 = \dfrac{1}{12} \dfrac{r^2 - 1}{r}$
$M_2 = M_1$		
$M_1^0 = -k_1^0 \, P \, l$	$k_1^0 = \dfrac{1}{8} \dfrac{r^2 - 1}{r}$	$k_1 = \dfrac{1}{8} \dfrac{r^2 - 1}{r}$
$A_0 = k_3 \, P$	$k_3 = \dfrac{1}{2}(r - 1)$	$k_3 = \dfrac{1}{2}(r - 1)$
$B_0 = A_0$		
$M_{0\,\max} = k_7 \, P \, l$	$k_7 = \dfrac{r}{8}$	$k_7 = \dfrac{1}{8} \dfrac{r^2 - 1}{r}$
$x_m = \dfrac{l}{2}$		
$\alpha_0 = k_9 \, P \, l^2$	$k_9 = \dfrac{1}{24} \dfrac{r^2 - 1}{r}$	$k_9 = \dfrac{1}{24} \dfrac{r^2 - 1}{r}$
$\beta_0 = \alpha_0$		

Table 28

r	k_1	k_1^0	k_3	k_7	k_9
1	0	0	0	0	0
2	0.125	0.188	0.500	0.250	0.061
3	0.222	0.333	1.000	0.333	0.111
4	0.313	0.469	1.500	0.500	0.156
5	0.400	0.600	2.000	0.600	0.200
6	0.486	0.729	2.500	0.750	0.243
7	0.571	0.857	3.000	0.857	0.286
8	0.656	0.984	3.500	1.000	0.328
9	0.741	1.111	4.000	1.111	0.370
10	0.825	1.238	4.500	1.250	0.413
11	0.909	1.364	5.000	1.364	0.455
12	0.993	1.490	5.500	1.500	0.497
13	1.077	1.615	6.000	1.615	0.538
14	1.161	1.741	6.500	1.750	0.580
15	1.244	1.867	7.000	1.867	0.622
16	1.328	1.992	7.500	2.000	0.664

Loading 29 (to Table 29)

Several symmetrically arranged concentrated loads, at distance a from each other, the first and the last at a distance $a/2$ from the supports.

$$l = r\,a$$

r loads P being present.

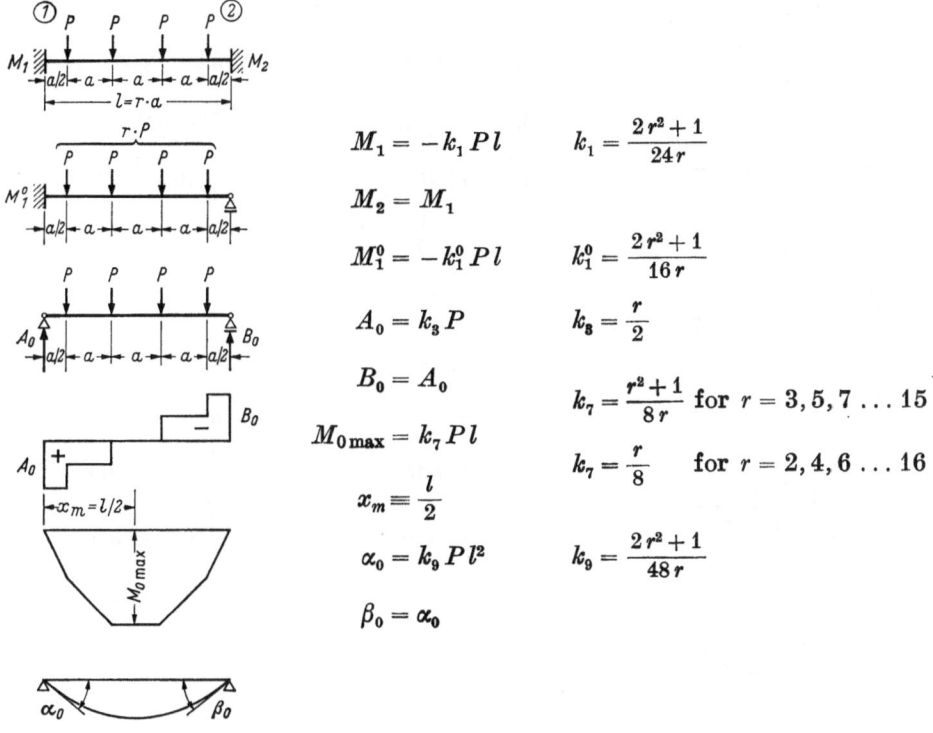

$$M_1 = -k_1\,P\,l \qquad k_1 = \frac{2\,r^2+1}{24\,r}$$

$$M_2 = M_1$$

$$M_1^0 = -k_1^0\,P\,l \qquad k_1^0 = \frac{2\,r^2+1}{16\,r}$$

$$A_0 = k_3\,P \qquad k_3 = \frac{r}{2}$$

$$B_0 = A_0$$

$$M_{0\,max} = k_7\,P\,l \qquad k_7 = \frac{r^2+1}{8\,r}\ \text{for}\ r = 3,5,7 \ldots 15$$

$$k_7 = \frac{r}{8}\qquad \text{for}\ r = 2,4,6 \ldots 16$$

$$x_m = \frac{l}{2}$$

$$\alpha_0 = k_9\,P\,l^2 \qquad k_9 = \frac{2\,r^2+1}{48\,r}$$

$$\beta_0 = \alpha_0$$

$$r = 3,5,7,9 \ldots k_7 = \frac{r^2+1}{8\,r} \qquad\qquad r = 2,4,6,8 \ldots k_7 = \frac{r}{8}$$

Table 29

r	k_1	k_1^0	k_3	k_7	k_9
1	0.125	0.188	0.500	0.250	0.063
2	0.188	0.281	1.000	0.250	0.094
3	0.264	0.396	1.500	0.417	0.132
4	0.344	0.516	2.000	0.500	0.172
5	0.425	0.638	2.500	0.650	0.213
6	0.507	0.760	3.000	0.750	0.253
7	0.589	0.884	3.500	0.893	0.295
8	0.672	1.008	4.000	1.000	0.336
9	0.755	1.132	4.500	1.139	0.377
10	0.838	1.256	5.000	1.250	0.419
11	0.920	1.381	5.500	1.386	0.460
12	1.003	1.505	6.000	1.500	0.502
13	1.087	1.630	6.500	1.635	0.543
14	1.170	1.754	7.000	1.750	0.585
15	1.253	1.879	7.500	1.883	0.626
16	1.336	2.004	8.000	2.000	0.668

Table 30   – 86 –

Loading 30 (to Table 30)

Application of a pure bending moment

M being positive when rotating in the represented direction

$$a = n\, l$$

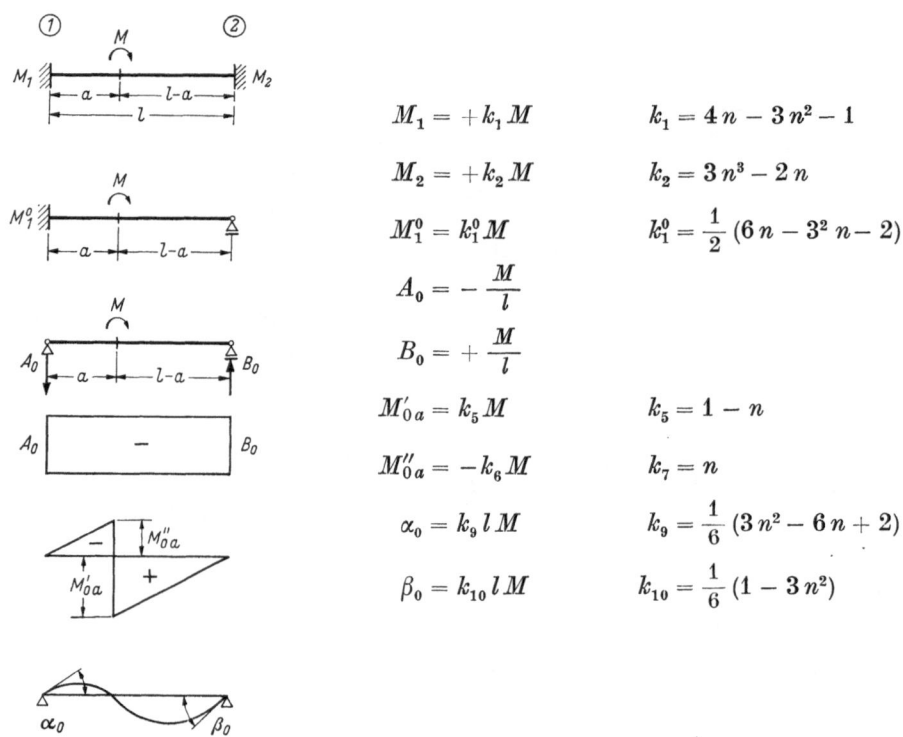

$$M_1 = +k_1\, M \qquad\qquad k_1 = 4\,n - 3\,n^2 - 1$$

$$M_2 = +k_2\, M \qquad\qquad k_2 = 3\,n^3 - 2\,n$$

$$M_1^0 = k_1^0\, M \qquad\qquad k_1^0 = \frac{1}{2}\,(6\,n - 3^2\,n - 2)$$

$$A_0 = -\,\frac{M}{l}$$

$$B_0 = +\,\frac{M}{l}$$

$$M'_{0a} = k_5\, M \qquad\qquad k_5 = 1 - n$$

$$M''_{0a} = -k_6\, M \qquad\qquad k_7 = n$$

$$\alpha_0 = k_9\, l\, M \qquad\qquad k_9 = \frac{1}{6}\,(3\,n^2 - 6\,n + 2)$$

$$\beta_0 = k_{10}\, l\, M \qquad\qquad k_{10} = \frac{1}{6}\,(1 - 3\,n^2)$$

Table 30

m	k_1	k_2	k_1^0	k_5	k_6	k_9	k_{10}	m
0.00	−1.000	0	−1.000	1.000	0	+0.333	+0.167	0.00
0.05	−0.808	−0.093	−0.854	0.950	0.050	+0.285	+0.165	0.05
0.10	−0.630	−0.170	−0.715	.900	.100	+0.238	+0.162	0.10
0.15	−0.468	−0.233	−0.584	.850	.150	+0.195	+0.155	0.15
0.20	−0.320	−0.280	−0.460	.800	.200	+0.153	+0.147	0.20
0.25	−0.188	−0.313	−0.344	.750	.250	+0.115	+0.135	0.25
0.30	−0.070	−0.330	−0.235	.700	.300	+0.078	+0.122	0.30
0.35	+0.033	−0.333	−0.134	.650	.350	+0.045	+0.105	0.35
0.40	+0.120	−0.320	−0.040	.600	.400	+0.013	+0.087	0.40
0.45	+0.193	−0.293	+0.046	.550	.450	−0.015	+0.065	0.45
0.50	+0.250	−0.250	+0.125	500	.500	−0.042	+0.042	0.50
0.55	+0.293	−0.193	+0.196	.450	.550	−0.065	+0.015	0.55
0.60	+0.320	−0.120	+0.260	.400	.600	−0.087	−0.013	0.60
0.65	+0.333	−0.033	+0.316	.350	.650	−0.105	−0.045	0.65
0.70	+0.330	+0.070	+0.365	.300	.700	−0.122	−0.078	0.70
0.75	+0.313	+0.188	+0.406	.250	.750	−0.135	−0.115	0.75
0.80	+0.280	+0.320	+0.440	.200	.800	−0.147	−0.153	0.80
0.85	+0.233	+.0468	+0.466	.150	.850	−0.155	−0.195	0.85
0.90	+0.170	+0.630	+0 485	.100	.900	−0.162	−0.238	0.90
0.95	+0.093	+0.808	+0.496	.050	.950	−0.165	−0.285	0.95
1.00	+0.000	+1.000	+0.500	0,000	1.000	−0.167	−0.333	1.00

Loading 31 (to Table 31)

$(r-1)$ equal bending moments in equal distances both
from each other and from the supports

$$l = ra$$

$i =$ ordinal number of the applied moments

Positive moments clockwise

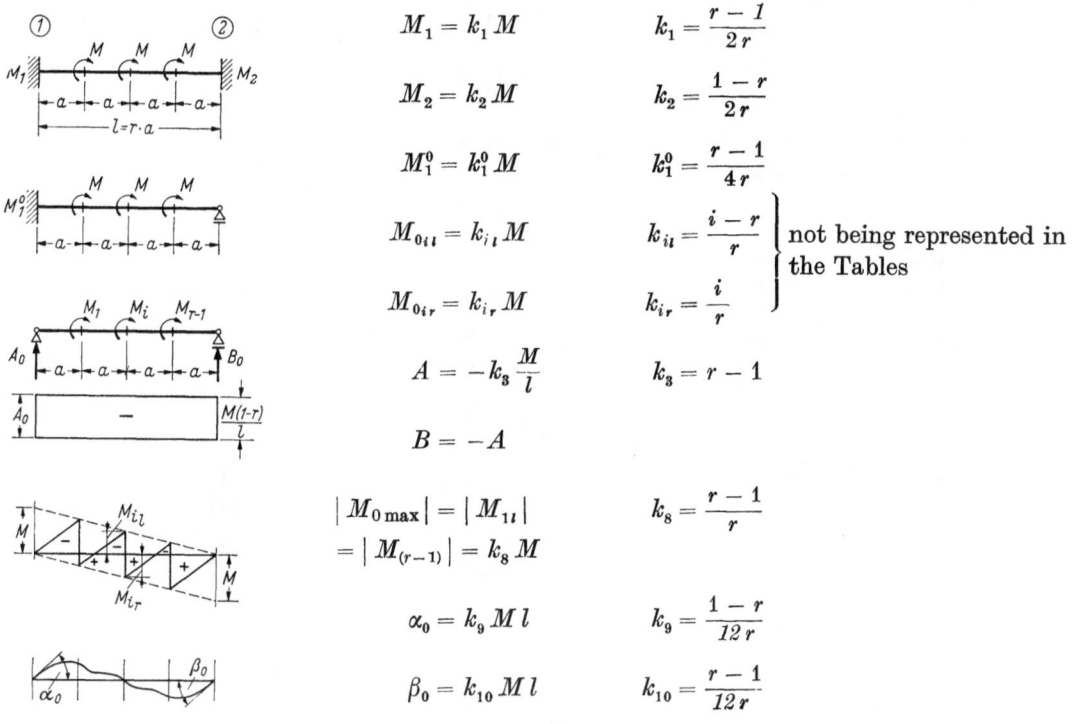

$$M_1 = k_1 M \qquad k_1 = \frac{r-1}{2r}$$

$$M_2 = k_2 M \qquad k_2 = \frac{1-r}{2r}$$

$$M_1^0 = k_1^0 M \qquad k_1^0 = \frac{r-1}{4r}$$

$$M_{0\,i\,l} = k_{i\,l} M \qquad k_{i\,l} = \frac{i-r}{r} \left.\right\} \begin{array}{l}\text{not being represented in}\\ \text{the Tables}\end{array}$$

$$M_{0\,i\,r} = k_{i\,r} M \qquad k_{i\,r} = \frac{i}{r}$$

$$A = -k_3 \frac{M}{l} \qquad k_3 = r-1$$

$$B = -A$$

$$|M_{0\max}| = |M_{1l}| \qquad k_8 = \frac{r-1}{r}$$
$$= |M_{(r-1)}| = k_8 M$$

$$\alpha_0 = k_9 M l \qquad k_9 = \frac{1-r}{12r}$$

$$\beta_0 = k_{10} M l \qquad k_{10} = \frac{r-1}{12r}$$

Table 31

r	k_1	k_2	k_1^0	k_3	k_8	k_9	k_{10}	r
1	0	0	0	0	0	0	0	1
2	0.250	−0.250	0.125	1.0	0.500	−0.042	0.042	2
3	0.333	−0.333	0.167	2.0	0.667	−0.056	0.056	3
4	0.375	−0.375	0.188	3.0	0.750	−0.063	0.063	4
5	0.400	−0.400	0.200	4.0	0.800	−0.067	0.067	5
6	0.417	−0.417	0.208	5.0	0.833	−0.069	0.069	6
7	0.429	−0.429	0.214	6.0	0.857	−0.071	0.071	7
8	0.438	−0.438	0.219	7.0	0.875	−0.073	0.073	8
9	0.444	−0.444	0.222	8.0	0.889	−0.074	0.074	9
10	0.450	−0.450	0.225	9.0	0.900	−0.075	0.075	10

Table 32   – 88 –

Loading 32 (to Table 32)

Application of r equal and equidistant (a) moments M, the first and the last at distance $a/2$ from the supports.

Positive moments clockwise.

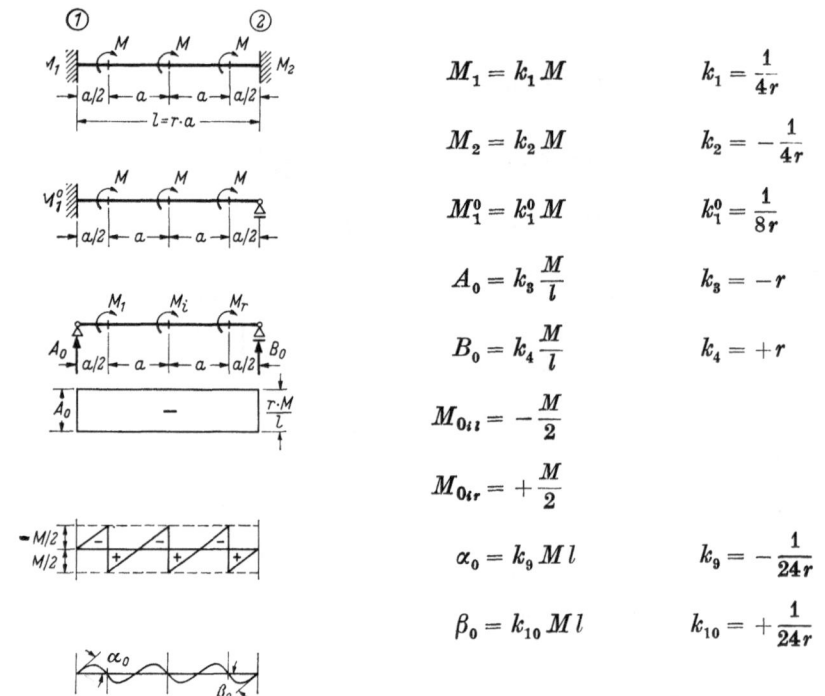

$$M_1 = k_1 M \qquad k_1 = \frac{1}{4r}$$

$$M_2 = k_2 M \qquad k_2 = -\frac{1}{4r}$$

$$M_1^0 = k_1^0 M \qquad k_1^0 = \frac{1}{8r}$$

$$A_0 = k_8 \frac{M}{l} \qquad k_8 = -r$$

$$B_0 = k_4 \frac{M}{l} \qquad k_4 = +r$$

$$M_{0il} = -\frac{M}{2}$$

$$M_{0ir} = +\frac{M}{2}$$

$$\alpha_0 = k_9 M l \qquad k_9 = -\frac{1}{24r}$$

$$\beta_0 = k_{10} M l \qquad k_{10} = +\frac{1}{24r}$$

Table 32

r	k_1	k_2	k_1^0	k_3	k_9	k_{10}	r
1	0.250	-0.250	0.125	-1	-0.042	0.042	1
2	0.125	-0.125	0.063	-2	-0.021	0.021	2
3	0.083	-0.083	0.042	-3	-0.014	0.014	3
4	0.063	-0.063	0.031	-4	-0.010	0.010	4
5	0.050	-0.050	0.025	-5	-0.008	0.008	5
6	0.042	-0.042	0.021	-6	-0.007	0.007	6
7	0.036	-0.036	0.018	-7	-0.006	0.006	7
8	0.031	-0.031	0.016	-8	-0.005	0.005	8
9	0.028	-0.028	0.014	-9	-0.005	0.005	9
10	0.025	-0.025	0.013	-10	-0.004	0.004	10

Loading 33

Settling of supports

Settling of the left support: δ_1

Settling of the right support: δ_2

$$\delta_2 - \delta_1 = \Delta\delta; \qquad \Delta\delta \text{ is an algebraical magnitude}$$

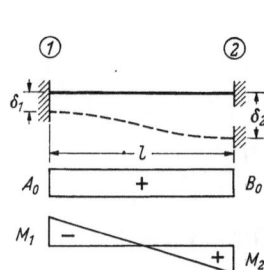

$$M_1 = -\frac{6\,EI\,\Delta\delta}{l^2}$$

$$M_2 = -M_1$$

$$M_1^0 = -\frac{3\,EI\,\Delta\delta}{l^2}$$

Reations at the two simple supports generally are:

$$A_0 = \frac{M_2 - M_1}{l}$$

$$B_0 = \frac{M_1 - M_2}{l} \qquad B_0 = -A_0$$

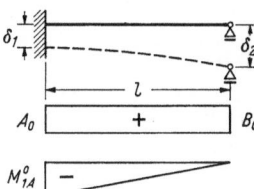

The equations developed for A_0 and B_0 are for

two built-in ends	one built-in end
$A_0 = +\dfrac{12\,EI\,\Delta\delta}{l^3}$	$A_0 = +\dfrac{3\,EI\,\Delta\delta}{l^3}$
$B_0 = -\dfrac{12\,EI\,\Delta\delta}{l^3}$	$B_0 = -\dfrac{3\,EI\,\Delta\delta}{l^3}$

Attention: For the case of settling of supports there are only these formulae. Tables would be unsuitable.

Loading 34

Differences in temperature

Assumptions:

1. Stationary heat flow $\Delta t^\circ = t_u^\circ - t_0^\circ$, Δt° being considered as an algebraical magnitude.

2. Modulus of elasticity E and moment of inertia I may be constant over the whole span.

$$\omega = \text{Linear thermal expansion}$$

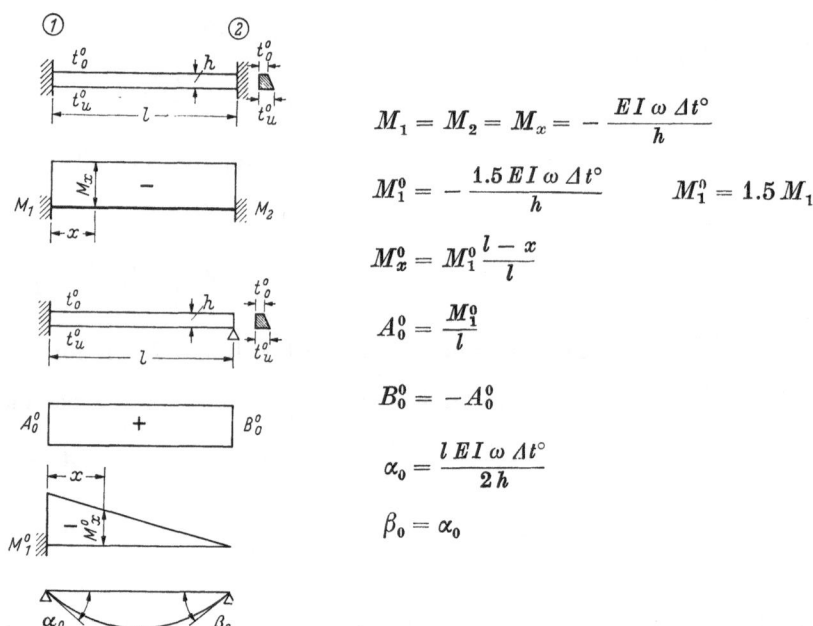

$$M_1 = M_2 = M_x = -\frac{EI\,\omega\,\Delta t^\circ}{h}$$

$$M_1^0 = -\frac{1.5\,EI\,\omega\,\Delta t^\circ}{h} \qquad M_1^0 = 1.5\,M_1$$

$$M_x^0 = M_1^0\,\frac{l-x}{l}$$

$$A_0^0 = \frac{M_1^0}{l}$$

$$B_0^0 = -A_0^0$$

$$\alpha_0 = \frac{l\,EI\,\omega\,\Delta t^\circ}{2\,h}$$

$$\beta_0 = \alpha_0$$

Attention: For the case of differences in temperature there are only these formulae. Tables would be unsuitable.

Table 35

Loads distributed over the whole span

Lastfall	k_1	k_2	k_1^0	k_3	k_4	k_7	k_8	k_9	k_{10}
	0.083	0.083	0.125	0.500	0.500	0.125	0.500	0.042	0.042
	.050	.033	.067	.333	.167	.064	.423	.022	.019
	.033	.050	.058	.167	.333	.064	.577	.019	.022
	.031	.031	.047	.250	.250	.042	.500	.016	.016
	.052	.052	.078	.250	.250	.083	.500	.026	.026
Quadr. parabola	.067	.067	.100	.333	.333	.104	.500	.033	.033
Quadr. parabola	.017	.017	.025	.167	.167	.021	.500	.008	.008
	.033	.017	.042	.250	.083	.043	.370	.014	.011
	.017	.033	.033	.083	.250	.043	.630	.011	.014
	.067	.050	.092	.417	.250	.086	.446	.031	.028
	.050	.067	.083	.250	.417	.086	.554	.028	.031
	.044	.044	.066	.250	.250	.063	.500	.022	.022
	0.039	0.039	0.059	0.250	0.250	0.063	0.500	0.020	0.020

The coefficients k correspond to those of loading 1 (see pp. 28 and 29).

Table V-1   – 92 –

Loading V-1 (to Table V-1)

(Prestressing)

Prestressing tendon according to a quadratic parabola, centrically anchored at the end.

$$e_x = \frac{e_m}{s^2}(2\,s\,x - x^2)$$

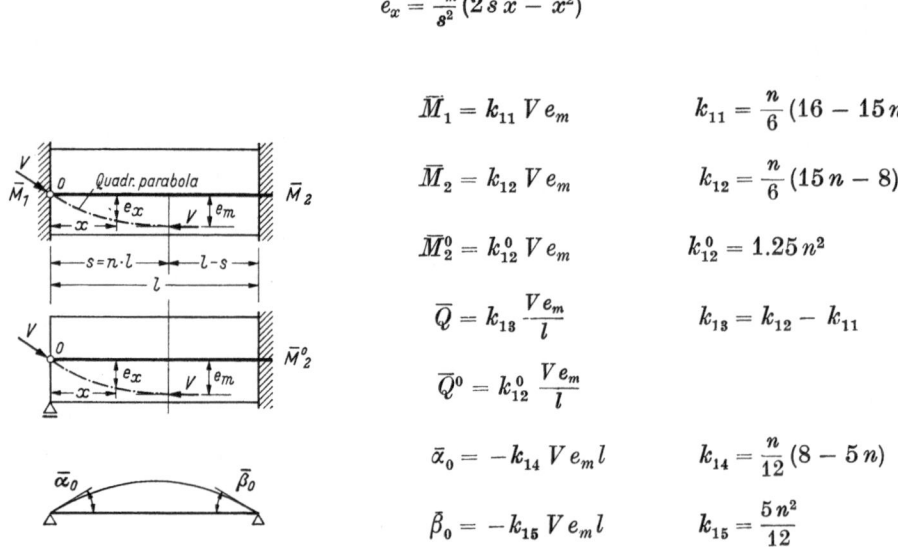

$$\bar{M}_1 = k_{11}\,V\,e_m \qquad\qquad k_{11} = \frac{n}{6}(16 - 15\,n)$$

$$\bar{M}_2 = k_{12}\,V\,e_m \qquad\qquad k_{12} = \frac{n}{6}(15\,n - 8)$$

$$\bar{M}_2^0 = k_{12}^0\,V\,e_m \qquad\qquad k_{12}^0 = 1.25\,n^2$$

$$\bar{Q} = k_{13}\frac{V\,e_m}{l} \qquad\qquad k_{13} = k_{12} - k_{11}$$

$$\bar{Q}^0 = k_{12}^0\frac{V\,e_m}{l}$$

$$\bar{\alpha}_0 = -k_{14}\,V\,e_m\,l \qquad\qquad k_{14} = \frac{n}{12}(8 - 5\,n)$$

$$\bar{\beta}_0 = -k_{15}\,V\,e_m\,l \qquad\qquad k_{15} = \frac{5\,n^2}{12}$$

Table V-1

n	k_{11}	k_{12}	k_{12}^0	k_{13}	k_{14}	k_{15}
0.050	0.127	−0.060	0.003	−0.188	0.032	0.001
0.100	0.242	−0.108	0.013	−0.350	0.063	0.004
0.150	0.344	−0.144	0.028	−0.488	0.091	0.009
0.200	0.433	−0.167	0.050	−0.600	0.117	0.017
0.250	0.511	−0.177	0.078	−0.688	0.141	0.026
0.300	0.575	−0.175	0.113	−0.750	0.163	0.038
0.325	0.603	−0.169	0.132	−0.772	0.173	0.044
0.350	0.627	−0.160	0.153	−0.788	0.182	0.051
0.375	0.648	−0.148	0.176	−0.797	0.191	0.059
0.400	0.667	−0.133	0.200	−0.800	0.200	0.067
0.425	0.682	−0.115	0.226	−0.797	0.208	0.075
0.450	0.694	−0.094	0.253	−0.788	0.216	0.084
0.475	0.703	−0.069	0.282	−0.772	0.223	0.094
0.500	0.708	−0.042	0.313	−0.750	0.229	0.104
0.525	0.711	−0.019	0.345	−0.730	0.235	0.115
0.550	0.710	+0.023	0.378	−0.688	0.241	0.126
0.575	0.707	+0.060	0.413	−0.647	0.246	0.138
0.600	0.700	+0.100	0.450	−0.600	0.250	0.150
0.625	0.690	+0.143	0.488	−0.547	0.254	0.163
0.650	0.677	+0.190	0.528	−0.488	0.257	0.176

Loading V-2 (to Table V-2)

(Prestressing)

Chapeau-Tendon

Prestressing tendon according to a quadratic parabola

$$e_2 + e_m = f$$
$$s = nl$$

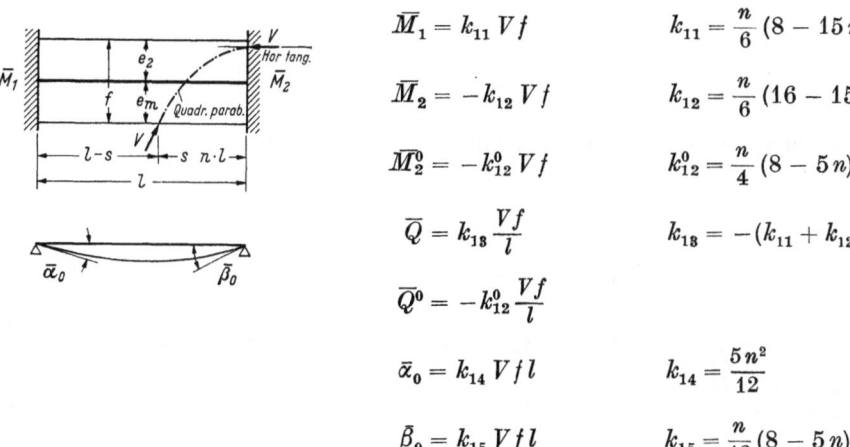

$$\overline{M}_1 = k_{11} V f \qquad k_{11} = \frac{n}{6} (8 - 15\,n)$$

$$\overline{M}_2 = -k_{12} V f \qquad k_{12} = \frac{n}{6} (16 - 15\,n)$$

$$\overline{M}_2^0 = -k_{12}^0 V f \qquad k_{12}^0 = \frac{n}{4} (8 - 5\,n)$$

$$\overline{Q} = k_{13} \frac{V f}{l} \qquad k_{13} = -(k_{11} + k_{12})$$

$$\overline{Q}^0 = -k_{12}^0 \frac{V f}{l}$$

$$\bar{\alpha}_0 = k_{14} V f l \qquad k_{14} = \frac{5\,n^2}{12}$$

$$\bar{\beta}_0 = k_{15} V f l \qquad k_{15} = \frac{n}{12} (8 - 5\,n)$$

Table V-2

n	k_{11}	k_{12}	k_{12}^0	k_{13}	k_{14}	k_{15}
0.200	0.167	0.433	0.350	−0.600	0.017	0.117
0.225	0.173	0.473	0.387	−0.647	0.021	0.129
0.250	0.177	0.511	0.422	−0.688	0.026	0.141
0.275	0.178	0.544	0.456	−0.633	0.032	0.152
0.300	0.175	0.575	0.488	−0.750	0.038	0.163
0.325	0.169	0.603	0.518	−0.772	0.044	0.173
0.350	0.160	0.627	0.547	−0.788	0.051	0.182
0.375	0.148	0.648	0.574	−0.797	0.059	0.191
0.400	0.133	0.667	0.600	−0.800	0.067	0.200
0.425	0.115	0.682	0.624	−0.797	0.075	0.208
0.450	0.094	0.694	0.647	−0.788	0.084	0.216
0.475	0.069	0.703	0.668	−0.772	0.094	0.223
0.500	0.042	0.708	0.688	−0.750	0.104	0.229

7*

Loading V-3 (to Table V-3 a and b)

(Prestressing)

Prestressing tendon according to a cubic parabola over intermediate **supports**

With $e_2 < 0$

$$s = n\,l$$

$$e_m = m\,(-e_2)$$

$$m = \frac{e_m}{-e_2}$$

and V negativ being a compression force, will be:

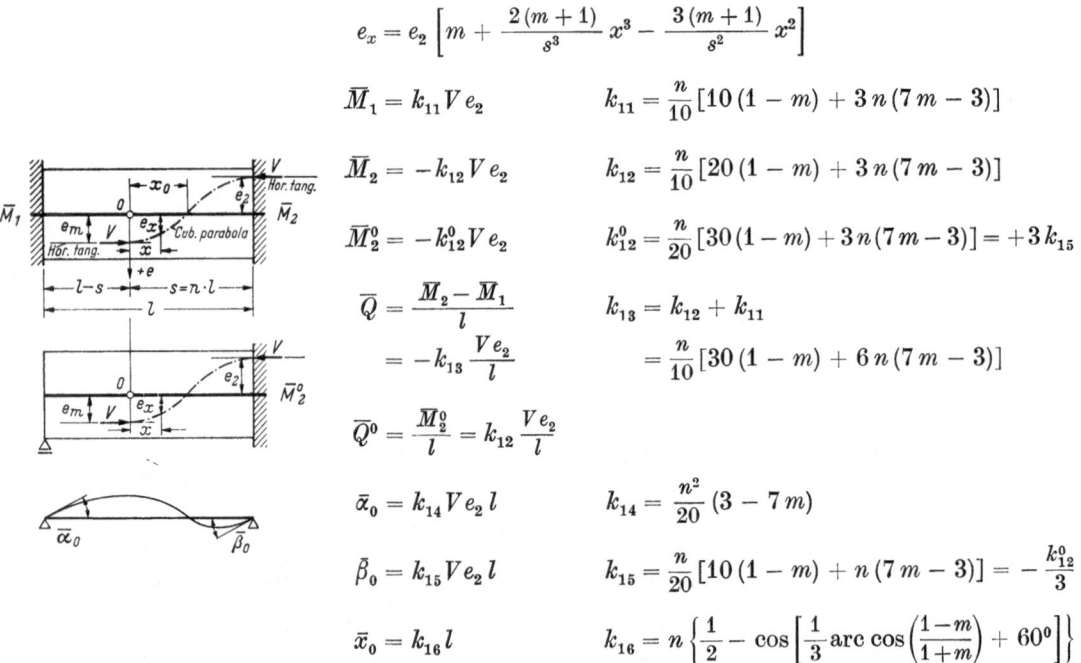

$$e_x = e_2 \left[m + \frac{2\,(m+1)}{s^3}\,x^3 - \frac{3\,(m+1)}{s^2}\,x^2 \right]$$

$$\overline{M}_1 = k_{11}\,V\,e_2 \qquad\qquad k_{11} = \frac{n}{10}\,[10\,(1-m) + 3\,n\,(7\,m-3)]$$

$$\overline{M}_2 = -k_{12}\,V\,e_2 \qquad\qquad k_{12} = \frac{n}{10}\,[20\,(1-m) + 3\,n\,(7\,m-3)]$$

$$\overline{M}_2^0 = -k_{12}^0\,V\,e_2 \qquad\qquad k_{12}^0 = \frac{n}{20}\,[30\,(1-m) + 3\,n\,(7\,m-3)] = +3\,k_{15}$$

$$\overline{Q} = \frac{\overline{M}_2 - \overline{M}_1}{l} \qquad\qquad k_{13} = k_{12} + k_{11}$$

$$\quad = -k_{13}\,\frac{V\,e_2}{l} \qquad\qquad\quad = \frac{n}{10}\,[30\,(1-m) + 6\,n\,(7\,m-3)]$$

$$\overline{Q}^0 = \frac{\overline{M}_2^0}{l} = k_{12}\,\frac{V\,e_2}{l}$$

$$\bar{\alpha}_0 = k_{14}\,V\,e_2\,l \qquad\qquad k_{14} = \frac{n^2}{20}\,(3 - 7\,m)$$

$$\bar{\beta}_0 = k_{15}\,V\,e_2\,l \qquad\qquad k_{15} = \frac{n}{20}\,[10\,(1-m) + n\,(7\,m-3)] = -\frac{k_{12}^0}{3}$$

$$\bar{x}_0 = k_{16}\,l \qquad\qquad k_{16} = n\left\{ \frac{1}{2} - \cos\left[\frac{1}{3}\,\text{arc cos}\left(\frac{1-m}{1+m}\right) + 60^0 \right] \right\}$$

For values k_{11}, k_{12} and k_{12}^0: see Table V-3 a

For values k_{14}, k_{15} and k_{16}: see Table V-3 b

In the above formulae V to be introduced by its absolute magnitude.

Table V-3a

Top number $= k_{11}$;   Medium number $= k_{12}$;   Bottom number $= k_{12}^0$

m	n												
	0.40	0.45	0.50	0.55	0.60	0.65	0.70	0.75	0.80	0.85	0.90	0.95	1.00
-0.10 $\begin{array}{c}k_{11}\\k_{12}\\k_{12}^0\end{array}$	0.262 / .702 / .571	0.270 / .765 / .630	0.273 / .823 / .686	0.269 / .874 / .740	0.260 / .920 / .790	0.246 / .961 / .838	0.226 / .996 / .883	0.201 / 1.206 / .925	0.170 / 1.050 / .965	0.133 / 1.068 / 1.002	0.091 / 1.081 / 1.036	0.043 / 1.088 / 1.067	-0.010 / 1.090 / 1.095
-0.05	.259 / .679 / .550	.269 / .742 / .607	.274 / .799 / .662	.274 / .851 / .714	.268 / .898 / .764	.258 / .940 / .811	.243 / .978 / .856	.222 / 1.010 / .899	.197 / 1.037 / .938	.166 / 1.059 / .976	.131 / 1.076 / 1.011	.091 / 1.088 / 1.043	.045 / 1.095 / 1.073
0	.256 / .656 / .528	.268 / .718 / .584	.275 / .775 / .638	.278 / .828 / .689	.276 / .876 / .738	.270 / .920 / .785	.259 / .959 / .830	.244 / .994 / .872	.224 / 1.024 / .912	.200 / 1.050 / .950	.171 / 1.071 / .986	.138 / 1.088 / 1.019	.100 / 1.100 / 1.050
0.05	.253 / .633 / .506	.267 / .694 / .561	.276 / .751 / .613	.282 / .805 / .664	.284 / .854 / .712	.282 / .899 / .758	.276 / .941 / .803	.265 / .978 / .845	.251 / 1.011 / .886	.233 / 1.041 / .924	.211 / 1.066 / .961	.185 / 1.088 / .995	.155 / 1.105 / 1.028
0.10	.250 / .610 / .485	.265 / .670 / .538	.278 / .728 / .589	.286 / .781 / .638	.292 / .832 / .686	.294 / .879 / .732	.292 / .922 / .776	.287 / .962 / .818	.278 / .998 / .859	.267 / 1.032 / .898	.251 / 1.061 / .936	.232 / 1.087 / .971	.210 / 1.110 / 1.005
0.15	.246 / .586 / .463	.264 / .647 / .515	.279 / .704 / .564	.291 / .758 / .613	.299 / .809 / .660	.305 / .858 / .705	.308 / .903 / .749	.308 / .946 / .792	.306 / .986 / .833	.300 / 1.022 / .872	.291 / 1.056 / .911	.280 / 1.087 / .947	.265 / 1.115 / .983
0.20	.243 / .563 / .442	.263 / .623 / .491	.280 / .680 / .540	.295 / .735 / .587	.307 / .787 / .634	.317 / .837 / .679	.325 / .885 / .722	.330 / .930 / .765	.333 / .973 / .806	.333 / 1.013 / .847	.331 / 1.051 / .886	.327 / 1.087 / .923	.320 / 1.120 / .960
0.25	.240 / .540 / .420	.262 / .599 / .468	.281 / .656 / .516	.299 / .712 / .562	.315 / .765 / .608	.329 / .817 / .652	.341 / .866 / .696	.352 / .914 / .738	.360 / .960 / .780	.367 / 1.004 / .821	.371 / 1.046 / .861	.374 / 1.087 / .900	.375 / 1.125 / .938
0.30	.237 / .517 / .398	.260 / .575 / .445	.283 / .633 / .491	.303 / .688 / .537	.323 / .743 / .581	.341 / .796 / .626	.358 / .848 / .669	.373 / .898 / .712	.387 / .947 / .754	.400 / .995 / .795	.411 / 1.041 / .836	.421 / 1.086 / .876	.430 / 1.130 / .915
0.35	.234 / .494 / .377	.259 / .552 / .422	.284 / .609 / .467	.308 / .665 / .511	.331 / .721 / .555	.353 / .775 / .599	.374 / .829 / .642	.395 / .882 / .685	.414 / .934 / .727	.433 / .986 / .769	.451 / 1.036 / .811	.469 / 1.086 / .852	.485 / 1.135 / .893
0.40	.230 / .470 / .355	.258 / .528 / .399	.285 / .585 / .443	.312 / .642 / .486	.338 / .698 / .529	.365 / .755 / .572	.391 / .811 / .615	.416 / .866 / .658	.442 / .922 / .701	.467 / .977 / .743	.491 / 1.031 / .786	.516 / 1.086 / .828	.540 / 1.140 / .870
0.45	.227 / .447 / .334	.257 / .504 / .376	.286 / .561 / .418	.316 / .619 / .461	.346 / .676 / .503	.377 / .734 / .546	.407 / .792 / .589	.438 / .850 / .631	.469 / .909 / .674	.500 / .968 / .718	.532 / 1.027 / .761	.563 / 1.086 / .804	.595 / 1.145 / .848
0.50	.224 / .424 / .312	.255 / .480 / .353	.288 / .538 / .394	.320 / .593 / .435	.354 / .654 / .477	.388 / .713 / .519	.424 / .774 / .562	.459 / .834 / .605	.496 / .896 / .648	.533 / .958 / .692	.572 / 1.022 / .736	.610 / 1.085 / .780	.650 / 1.150 / .825
0.55	.221 / .401 / .290	.254 / .457 / .330	.289 / .514 / .369	.325 / .572 / .410	.362 / .632 / .451	.400 / .693 / .493	.440 / .755 / .535	.481 / .818 / .578	.523 / .883 / .622	.567 / .949 / .666	.612 / 1.017 / .711	.658 / 1.085 / .756	.705 / 1.155 / .803
0.60	.218 / .378 / .269	.253 / .433 / .307	.290 / .490 / .345	.329 / .549 / .385	.370 / .610 / .425	.412 / .672 / .466	.456 / .736 / .508	.503 / .803 / .551	.550 / .870 / .595	.600 / .940 / .640	.652 / 1.012 / .686	.705 / 1.085 / .733	.760 / 1.160 / .780
0.65	.214 / .354 / .247	.252 / .409 / .283	.291 / .466 / .321	.333 / .526 / .359	.377 / .587 / .399	.424 / .652 / .440	.473 / .718 / .481	.524 / .787 / .525	.578 / .858 / .569	.634 / .931 / .614	.692 / 1.007 / .661	.752 / 1.085 / .709	.815 / 1.165 / .758

Table V-3 b — 96 —

Table V-3a (Continuation)

m		n												
		0.40	0.45	0.50	0.55	0.60	0.65	0.70	0.75	0.80	0.85	0.90	0.95	1.00
0.70	k_{11}	.211	.250	.293	.337	.385	.436	.489	.546	.605	.667	.732	.799	.870
	k_{12}	.331	.385	.443	.502	.565	.631	.699	.771	.845	.922	1.002	1.084	1.170
	k_{12}^{0}	.226	.260	.296	.334	.373	.413	.455	.498	.542	.588	.636	.685	.735
0.75	k_{11}	.208	.249	.294	.342	.393	.448	.506	.567	.632	.700	.772	.847	.925
	k_{12}	.308	.362	.419	.479	.543	.610	.681	.755	.832	.913	.997	1.084	1.175
	k_{12}^{0}	.204	.237	.272	.308	.347	.386	.428	.471	.516	.563	.611	.661	.713
0.80	k_{11}	.205	.248	.295	.346	.401	.460	.522	.589	.659	.734	.812	.894	.980
	k_{12}	.285	.338	.395	.456	.521	.590	.662	.739	.819	.904	.992	1.084	1.180
	k_{12}^{0}	.182	.214	.248	.283	.320	.360	.401	.444	.490	.537	.586	.637	.690
0.85	k_{11}	.202	.247	.296	.350	.409	.471	.539	.610	.686	.767	.852	.941	1.035
	k_{12}	.262	.314	.371	.433	.499	.569	.644	.723	.806	.894	.987	1.084	1.185
	k_{12}^{0}	.161	.191	.223	.278	.294	.333	.374	.418	.463	.511	.561	.613	.668
0.90	k_{11}	.198	.246	.298	.355	.416	.483	.555	.632	.714	.800	.892	.989	1.090
	k_{12}	.238	.291	.348	.410	.476	.548	.625	.707	.794	.885	.982	1.084	1.190
	k_{12}^{0}	.139	.168	.199	.232	.268	.307	.348	.391	.437	.485	.536	.589	.645
0.95	k_{11}	.195	.244	.299	.359	.424	.495	.572	.653	.741	.834	.932	1.036	1.145
	k_{12}	.215	.267	.324	.386	.454	.528	.607	.691	.781	.876	.977	1.083	1.195
	k_{12}^{0}	.118	.145	.174	.207	.242	.280	.321	.364	.410	.459	.511	.565	.623
1.00	k_{11}	.192	.243	.300	.363	.432	.507	.588	.675	.768	.867	.972	1.083	1.200
	k_{12}	.192	.243	.300	.363	.432	.507	.588	.675	.768	.867	.972	1.083	1.200
	k_{12}^{0}	.096	.122	.150	.182	.216	.254	.294	.338	.384	.434	.486	.542	.600

Table V-3b

Top number $= k_{14}$;   Medium number $= k_{15}$;   Bottom number $= k_{16}$

m		n												
		0.40	0.45	0.50	0.55	0.60	0.65	0.70	0.75	0.80	0.85	0.90	0.95	1.00
-0.10	k_{14}	0.030	0.038	0.046	0.056	0.067	0.078	0.091	0.104	0.118	0.134	0.150	0.167	0.185
	k_{15}	.190	.210	.229	.247	.263	.279	.294	.308	.322	.334	.345	.356	.365
	k_{16}	–	–	–	–	–	–	–	–	–	–	–	–	–
-0.05	k_{14}	.027	.034	.042	.051	.060	.071	.082	.094	.107	.121	.136	.151	.168
	k_{15}	.183	.202	.221	.238	.255	.271	.285	.300	.313	.325	.337	.348	.358
	k_{16}	–	–	–	–	–	–	–	–	–	–	–	–	–
0	k_{14}	.024	.030	.038	.045	.054	.063	.074	.084	.096	.108	.122	.135	.150
	k_{15}	.176	.195	.213	.230	.246	.262	.277	.291	.304	.317	.329	.340	.350
	k_{16}	–	–	–	–	–	–	–	–	–	–	–	–	–
0.05	k_{14}	.021	.027	.033	.040	.048	.056	.065	.075	.085	.096	.107	.120	.133
	k_{15}	.169	.187	.204	.221	.237	.253	.268	.282	.295	.308	.320	.332	.343
	k_{16}	.053	.059	.066	.073	.079	.086	.092	.099	.106	.112	.119	.125	.132
0.10	k_{14}	.018	.023	.029	.035	.041	.049	.056	.065	.074	.083	.093	.104	.115
	k_{15}	.162	.179	.196	.213	.229	.244	.259	.273	.286	.299	.312	.324	.335
	k_{16}	.074	.084	.093	.102	.112	.121	.130	.140	.149	.158	.167	.177	.186
0.15	k_{14}	.016	.020	.024	.030	.035	.041	.048	.055	.062	.070	.079	.088	.098
	k_{15}	.154	.172	.188	.204	.220	.235	.250	.264	.278	.291	.304	.316	.328
	k_{16}	.090	.102	.113	.124	.135	.147	.158	.169	.181	.192	.203	.215	.226
0.20	k_{14}	.013	.016	.020	.024	.029	.034	.039	.045	.051	.058	.065	.072	.080
	k_{15}	.147	.164	.188	.196	.211	.226	.241	.255	.269	.282	.295	.308	.320
	k_{16}	.104	.117	.130	.143	.156	.169	.181	.194	.207	.220	.233	.246	.259

Table V-3 b (Continuation)

m		n												
		0.40	0.45	0.50	0.55	0.60	0.65	0.70	0.75	0.80	0.85	0.90	0.95	1.00
0.25	k_{14}	.010	.013	.016	.019	.023	.026	.031	.035	.040	.045	.051	.056	.063
	k_{15}	.140	.156	.172	.187	.203	.217	.232	.246	.260	.274	.287	.300	.313
	k_{16}	.115	.129	.144	.158	.172	.187	.201	.215	.230	.244	.258	.273	.287
0.30	k_{14}	.007	.009	.011	.014	.016	.019	.022	.025	.029	.033	.037	.041	.045
	k_{15}	.133	.148	.164	.179	.194	.209	.223	.237	.251	.265	.279	.292	.305
	k_{16}	.125	.140	.156	.171	.187	.203	.218	.234	.249	.265	.280	.296	.312
0.35	k_{14}	.004	.006	.007	.008	.010	.012	.014	.015	.018	.020	.022	.025	.028
	k_{15}	.126	.141	.156	.170	.185	.200	.214	.228	.242	.256	.270	.284	.298
	k_{16}	.133	.150	.167	.183	.200	.217	.233	.250	.267	.283	.300	.317	.333
0.40	k_{14}	.002	.002	.003	.003	.004	.004	.005	.006	.006	.007	.008	.009	.010
	k_{15}	.118	.133	.148	.162	.176	.191	.205	.219	.234	.248	.262	.276	.290
	k_{16}	.141	.159	.177	.194	.212	.230	.247	.265	.283	.300	.318	.336	.353
0.45	k_{14}	-0.001	-0.002	-0.002	-0.002	-0.003	-0.003	-0.004	-0.004	-0.005	-0.005	-0.006	-0.007	-0.008
	k_{15}	.111	.125	.139	.154	.168	.182	.196	.211	.225	.239	.254	.268	.283
	k_{16}	.148	.167	.185	.204	.222	.241	.260	.278	.297	.315	.334	.352	.371
0.50	k_{14}	-0.004	-0.005	-0.006	-0.008	-0.009	-0.011	-0.012	-0.014	-0.016	-0.018	-0.020	-0.023	-0.025
	k_{15}	.104	.118	.131	.145	.159	.173	.187	.202	.216	.231	.245	.260	.275
	k_{16}	.155	.174	.194	.213	.232	.252	.271	.291	.310	.329	.349	.368	.387
0.55	k_{14}	-0.007	-0.009	-0.011	-0.013	-0.015	-0.018	-0.021	-0.024	-0.027	-0.031	-0.034	-0.038	-0.043
	k_{15}	.097	.110	.123	.137	.150	.164	.178	.193	.207	.222	.237	.252	.268
	k_{16}	.161	.181	.201	.221	.241	.261	.281	.301	.322	.342	.362	.382	.402
0.60	k_{14}	-0.010	-0.012	-0.015	-0.018	-0.022	-0.025	-0.029	-0.034	-0.038	-0.043	-0.049	-0.054	-0.060
	k_{15}	.090	.102	.115	.128	.142	.155	.169	.184	.198	.213	.229	.244	.260
	k_{16}	.166	.187	.208	.229	.250	.270	.291	.312	.333	.354	.374	.395	.416
0.65	k_{14}	-0.012	-0.016	-0.019	-0.023	-0.028	-0.033	-0.038	-0.044	-0.050	-0.056	-0.063	-0.070	-0.078
	k_{15}	.082	.094	.107	.120	.133	.147	.161	.175	.190	.205	.220	.236	.253
	k_{16}	.170	.192	.213	.234	.255	.277	.298	.319	.341	.362	.383	.404	.426
0.70	k_{14}	-0.015	-0.019	-0.024	-0.029	-0.034	-0.040	-0.047	-0.053	-0.061	-0.069	-0.077	-0.086	-0.095
	k_{15}	.075	.087	.099	.111	.124	.138	.152	.166	.181	.196	.212	.228	.245
	k_{16}	.173	.195	.217	.238	.260	.282	.303	.325	.347	.368	.390	.412	.433
0.75	k_{14}	-0.018	-0.023	-0.028	-0.034	-0.041	-0.048	-0.055	-0.063	-0.072	-0.081	-0.091	-0.102	-0.113
	k_{15}	.068	.079	.091	.103	.116	.129	.143	.157	.172	.188	.204	.220	.238
	k_{16}	.181	.204	.226	.249	.271	.294	.317	.339	.362	.384	.407	.430	.452
0.80	k_{14}	-0.021	-0.026	-0.033	-0.039	-0.047	-0.055	-0.064	-0.073	-0.083	-0.094	-0.105	-0.117	-0.130
	k_{15}	.061	.071	.083	.094	.107	.120	.134	.148	.163	.179	.195	.212	.230
	k_{16}	.185	.208	.231	.255	.278	.301	.324	.347	.370	.393	.417	.440	.463
0.85	k_{14}	-0.024	-0.030	-0.037	-0.045	-0.053	-0.062	-0.072	-0.083	-0.094	-0.107	-0.120	-0.133	-0.148
	k_{15}	.054	.064	.074	.086	.098	.111	.125	.139	.154	.170	.187	.204	.223
	k_{16}	.189	.213	.237	.260	.284	.307	.331	.355	.378	.402	.426	.449	.473
0.90	k_{14}	-0.026	-0.033	-0.041	-0.050	-0.059	-0.070	-0.081	-0.093	-0.106	-0.119	-0.134	-0.149	-0.165
	k_{15}	.046	.056	.066	.077	.089	.102	.116	.130	.146	.162	.179	.196	.215
	k_{16}	.193	.217	.241	.265	.290	.314	.338	.362	.386	.410	.434	.458	.483
0.95	k_{14}	-0.029	-0.037	-0.046	-0.055	-0.066	-0.077	-0.089	-0.103	-0.117	-0.132	-0.148	-0.165	-0.183
	k_{15}	.039	.048	.058	.069	.081	.093	.107	.121	.137	.153	.170	.189	.208
	k_{16}	.197	.221	.246	.270	.295	.320	.344	.369	.393	.418	.442	.467	.492
1.00	k_{14}	-0.032	-0.041	-0.050	-0.061	-0.072	-0.085	-0.098	-0.113	-0.128	-0.145	-0.162	-0.181	-0.200
	k_{15}	.032	.041	.050	.061	.072	.085	.098	.113	.128	.145	.162	.181	.200
	k_{16}	.200	.225	.250	.275	.300	.325	.350	.375	.400	.425	.450	.475	.500

Table V-4   – 98 –

Loading V-4 (to Table V-4)

(Prestressing)

Combination in end-span: Prestressing tendon according to a quadratic parabola changing into cubic parabola.

Quadratic parabola: $\quad l(n-1) \le x \le 0 \qquad$ for $e_x = f(x)$ see p. 92

Cubic parabola: $\quad\quad\quad 0 \le x \le nl \qquad$ for $e_x = f(x)$ see p. 94

$$e_2 < 0$$

$$m = \frac{e_m}{-e_2}$$

$$V < 0 \quad \text{(Prestressing force} = V)$$

In the following formulae the prestressing force V is to be introduced by its absolute value because its compression quality has already been considered in the formulae.

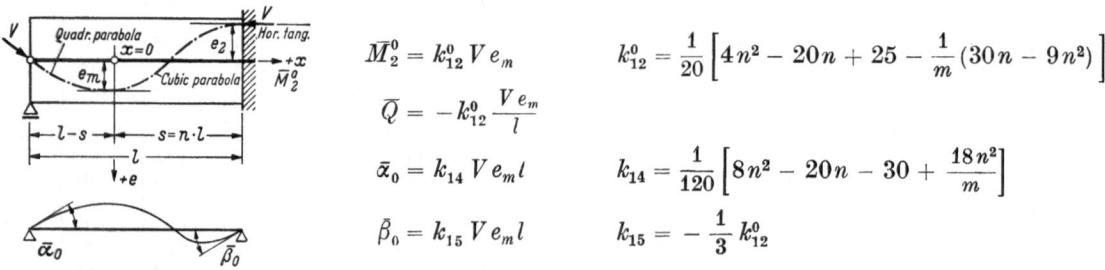

$$\bar{M}_2^0 = k_{12}^0 V e_m \qquad\qquad k_{12}^0 = \frac{1}{20}\left[4n^2 - 20n + 25 - \frac{1}{m}(30n - 9n^2)\right]$$

$$\bar{Q} = - k_{12}^0 \frac{V e_m}{l}$$

$$\bar{\alpha}_0 = k_{14} V e_m l \qquad\qquad k_{14} = \frac{1}{120}\left[8n^2 - 20n - 30 + \frac{18n^2}{m}\right]$$

$$\bar{\beta}_0 = k_{15} V e_m l \qquad\qquad k_{15} = -\frac{1}{3} k_{12}^0$$

The formulae for the coefficients k are combinations of those given under V-1 and V-3. So no special formulae have been developed for the coefficients k.

Table V-4

Top number k_{12}^0;   Medium number $= k_{14}$;   Bottom number $= k_{15}$

m		n											m
		0.40	0.45	0.50	0.55	0.60	0.65	0.70	0.75	0.80	0.85	0.90	
0.20	k_{12}^0 k_{14} k_{15}								− 3.747 +0.084 +1.249	− 3.982 +0.139 +1.327	− 4.205 +0.198 +1.402	− 4.416 +0.262 +1.472	k_{12}^0 k_{14} 0.20 k_{15}
0.25	k_{12}^0 k_{14} k_{15}							− 2.670 − 0.040 +0.890	− 2.875 +0.000 +0.958	− 3.070 +0.043 +1.023	− 3.255 +0.090 +1.085	− 3.430 +0.140 +1.143	k_{12}^0 k_{14} 0.25 k_{15}
0.30	k_{12}^0 k_{14} k_{15}						− 1.932 − 0.119 +0.644	− 2.117 − 0.089 +0.706	− 2.294 − 0.056 +0.764	− 2.462 − 0.021 +0.821	− 2.622 +0.018 +0.874	− 2.773 +0.059 +0.925	k_{12}^0 k_{14} 0.30 k_{15}
0.35	k_{12}^0 k_{14} k_{15}				− 1.387 − 0.172 +0.462	− 1.558 − 0.149 +0.519	− 1.722 − 0.124 +0.574	− 1.879 − 0.097 +0.626	− 2.028 − 0.066 +0.676	− 2.169 − 0.034 +0.723			k_{12}^0 k_{14} 0.35 k_{15}
0.40	k_{12}^0 k_{14} k_{15}				− 0.962 − 0.208 +0.321	− 1.123 − 0.191 +0.374	− 1.278 − 0.172 +0.426	− 1.426 − 0.150 +0.475	− 1.567 − 0.127 +0.523	− 1.702 − 0.101 +0.567	− 1.830 − 0.073 +0.610		k_{12}^0 k_{14} 0.40 k_{15}

Table V-4 (Continuation)

m		n 0.40	0 45	0.50	0.55	0.60	0.65	0.70	0.75	0.80	0.85	0.90		m
0.45	k_{12}^0			− 0.617	− 0.771	− 0.918	− 1.060	− 1.195	− 1.325	− 1.449			k_{12}^0	
	k_{14}			− 0.233	− 0.221	− 0.206	− 0.189	− 0.171	− 0.150	− 0.127			k_{14}	0.45
	k_{15}			+ 0.206	+ 0.257	+ 0.306	+ 0.353	+ 0.399	+ 0.442	+ 0.483			k_{15}	
0.50	k_{12}^0		− 0.327	− 0.475	− 0.617	− 0.754	− 0.885	− 1.011	− 1.131	− 1.246			k_{12}^0	
	k_{14}		− 0.251	− 0.242	− 0.231	− 0.218	− 0.204	− 0.187	− 0.169	− 0.149			k_{14}	0.50
	k_{15}		+ 0.109	+ 0.158	+ 0.206	+ 0.251	+ 0.295	+ 0.337	+ 0.377	+ 0.415			k_{15}	
0.55	k_{12}^0	− 0.078	− 0.221	− 0.359	− 0.492	− 0.620	− 0.743	− 0.860	− 0.973				k_{12}^0	
	k_{14}	− 0.262	− 0.256	− 0.249	− 0.239	− 0.228	− 0.215	− 0.200	− 0.184				k_{14}	0.55
	k_{15}	+ 0.026	+ 0.074	+ 0.120	+ 0.164	+ 0.207	+ 0.248	+ 0.287	+ 0.324				k_{15}	
0.60	k_{12}^0	+ 0.002	− 0.133	− 0.263	− 0.388	− 0.508	− 0.624	− 0.735	− 0.841				k_{12}^0	
	k_{14}	− 0.266	− 0.261	− 0.254	− 0.246	− 0.236	− 0.225	− 0.212	− 0.196				k_{14}	0.60
	k_{15}	− 0.001	+ 0.044	+ 0.088	+ 0.129	+ 0.169	+ 0.208	+ 0.245	+ 0.280				k_{15}	
0.65	k_{12}^0	+ 0.070	− 0.058	− 0.181	− 0.299	− 0.413	− 0.523	− 0.628					k_{12}^0	
	k_{14}	− 0.269	− 0.265	− 0.259	− 0.252	− 0.243	− 0.233	− 0.221					k_{14}	0.65
	k_{15}	− 0.023	+ 0.019	+ 0.060	+ 0.100	+ 0.138	+ 0.188	+ 0.209					k_{15}	
0.70	k_{12}^0	+ 0.128	+ 0.006	− 0.111	− 0.224	− 0.332	− 0.437	− 0.537					k_{12}^0	
	k_{14}	− 0.272	− 0.268	− 0.263	− 0.257	− 0.249	− 0.230	− 0.229					k_{14}	0.70
	k_{15}	− 0.043	− 0.002	+ 0.037	+ 0.075	+ 0.111	+ 0.146	+ 0.179					k_{15}	
0.75	k_{12}^0	+ 0.178	+ 0.062	− 0.050	− 0.158	− 0.262	− 0.362						k_{12}^0	
	k_{14}	− 0.274	− 0.271	− 0.267	− 0.261	− 0.254	− 0.246						k_{14}	0.75
	k_{15}	− 0.059	− 0.021	+ 0.017	+ 0.053	+ 0.087	+ 0.121						k_{15}	
0.80	k_{12}^0	+ 0.222	+ 0.111	+ 0.003	− 0.101	− 0.201	− 0.297						k_{12}^0	
	k_{14}	− 0.276	− 0.274	− 0.270	− 0.265	− 0.259	− 0.251						k_{14}	0.80
	k_{15}	− 0.074	− 0.037	− 0.001	+ 0.034	+ 0.067	+ 0.099						k_{15}	
0.85	k_{12}^0	+ 0.261	+ 0.154	+ 0.050	− 0.074	− 0.146	− 0.239						k_{12}^0	
	k_{14}	− 0.278	− 0.276	− 0.273	− 0.268	− 0.263	− 0.256						k_{14}	0.85
	k_{15}	− 0.087	− 0.051	− 0.017	+ 0.017	+ 0.049	+ 0.080						k_{15}	
0.90	k_{12}^0	+ 0.295	+ 0.192	+ 0.112	− 0.005	− 0.098	− 0.188						k_{12}^0	
	k_{14}	− 0.279	− 0.278	− 0.275	− 0.271	− 0.266	− 0.260						k_{14}	0.90
	k_{15}	− 0.098	− 0.064	− 0.031	+ 0.002	+ 0.033	+ 0.063						k_{15}	
0.95	k_{12}^0	+ 0.326	+ 0.226	+ 0.129	+ 0.035	− 0.055	− 0.142						k_{12}^0	
	k_{14}	− 0.281	− 0.280	− 0.277	− 0.274	− 0.269	− 0.264						k_{14}	0.95
	k_{15}	− 0.109	− 0.075	− 0.043	− 0.012	+ 0.018	+ 0.047						k_{15}	
1.00	k_{12}^0	+ 0.354	+ 0.257	+ 0.163	+ 0.072	− 0.016	− 0.100						k_{12}^0	
	k_{14}	− 0.282	− 0.281	− 0.279	− 0.276	− 0.272	− 0.267						k_{14}	1.00
	k_{15}	− 0.118	− 0.086	− 0.054	− 0.024	+ 0.005	+ 0.034						k_{15}	

Loading V-5 (to Tables V-5a–k)

(Prestressing)

Prestressing: Tendon over a whole inner span according to a cubic parabola.

In the following formulae the prestressing force V is to be introduced by its absolute value. The compression quality of V has already been considered.

$$e_1 : e_2 = r$$
$$\frac{e_m}{-e_2} = m$$
$$s = nl$$

$$\bar{M}_1 = k_{11} V e_2 \qquad k_{11} = \frac{1}{10}\left[1 + 9n^2(r-1) + 4n(2-5r) + m(11-12n)\right]$$

$$\bar{M}_2 = k_{12} V e_2 \qquad k_{12} = \frac{1}{10}\left[9n^2(1-r) + 2n(1+5r) + m(12n-1) - 11\right]$$

$$\bar{Q} = k_{13} \frac{V e_2}{l} \qquad k_{13} = \frac{6}{10}\left[3n^2(1-r) + n(5r-1) + 2m(2n-1) - 2\right]$$

$$\bar{\alpha}_0 = k_{14} V e_2 l \qquad k_{14} = \frac{1}{20}\left[3n^2(1-r) + 2n(5r-3) + m(4n-7) + 3\right]$$

$$\bar{\beta}_0 = k_{15} V e_2 l \qquad k_{15} = \frac{1}{20}\left[3n^2(r-1) - 4n(1+m) - 3m + 7\right]$$

Arranged according to parameter r the above mentioned coefficients k are found as follows:

$$k_{11}, k_{12} \text{ und } k_{13} \text{ in den Tab. V-5a bis V-5e}$$
$$k_{14} \text{ und } k_{15} \text{ in den Tab. V-5f bis V-5k}$$

For the equation of prestressing tendon axis $e_x = f(x)$ see p. 94.

Table V-5a

Top number $= k_{11}$;   Medium number $= k_{12}$;   Bottom number $= k_{13}$

m		$r = 0.55$					$r = 0.60$					m
		$n=0.40$	$n=0.45$	$n=0.50$	$n=0.55$	$n=0.60$	$n=0.40$	$n=0.45$	$n=0.50$	$n=0.55$	$n=0.60$	
-0.10	k_{11}	-0.147	-0.173	-0.201	-0.232	-0.264	-0.180	-0.209	-0.240	-0.273	-0.308	k_{11} -0.10
	k_{12}	-0.773	-0.725	-0.674	-0.621	-0.566	-0.760	-0.711	-0.660	-0.607	-0.552	k_{12}
	k_{13}	-0.626	-0.552	-0.473	-0.390	-0.302	-0.581	-0.502	-0.420	-0.334	-0.245	k_{13}
-0.05	k_{11}	-0.116	-0.145	-0.176	-0.210	-0.245	-0.149	-0.181	-0.215	-0.251	-0.289	k_{11} -0.05
	k_{12}	-0.754	-0.703	-0.649	-0.593	-0.535	-0.741	-0.689	-0.635	-0.579	-0.521	k_{12}
	k_{13}	-0.638	-0.558	-0.473	-0.384	-0.290	-0.593	-0.508	-0.420	-0.328	-0.233	k_{13}
0	k_{11}	-0.085	-0.117	-0.151	-0.188	-0.226	-0.118	-0.153	-0.190	-0.229	-0.270	k_{11} 0
	k_{12}	-0.735	-0.681	-0.624	-0.565	-0.504	-0.722	-0.667	-0.610	-0.551	-0.490	k_{12}
	k_{13}	-0.650	-0.564	-0.473	-0.378	-0.278	-0.605	-0.514	-0.420	-0.322	-0.221	k_{13}
0.05	k_{11}	-0.054	-0.089	-0.126	-0.166	-0.207	-0.087	-0.125	-0.165	-0.207	-0.250	k_{11} 0.05
	k_{12}	-0.716	-0.659	-0.599	-0.537	-0.473	-0.703	-0.645	-0.585	-0.523	-0.459	k_{12}
	k_{13}	-0.662	-0.570	-0.473	-0.372	-0.266	-0.617	-0.520	-0.420	-0.316	-0.209	k_{13}
0.10	k_{11}	-0.023	-0.061	-0.101	-0.144	-0.188	-0.056	-0.097	-0.140	-0.185	-0.232	k_{11} 0.10
	k_{12}	-0.697	-0.637	-0.574	-0.509	-0.442	-0.684	-0.623	-0.560	-0.495	-0.428	k_{12}
	k_{13}	-0.674	-0.576	-0.473	-0.366	-0.254	-0.629	-0.526	-0.420	-0.310	-0.197	k_{13}
0.15	k_{11}	0.008	-0.033	-0.076	-0.122	-0.169	-0.025	-0.069	-0.115	-0.163	-0.213	k_{11} 0.15
	k_{12}	-0.678	-0.615	-0.549	-0.481	-0.411	-0.665	-0.601	-0.535	-0.467	-0.397	k_{13}
	k_{13}	-0.686	-0.582	-0.473	-0.360	-0.242	-0.641	-0.532	-0.420	-0.304	-0.185	k_{13}
0.20	k_{11}	0.039	-0.005	-0.051	-0.100	-0.150	0.006	-0.041	-0.090	-0.141	-0.194	k_{11} 0.20
	k_{12}	-0.659	-0.593	-0.524	-0.453	-0.380	-0.646	-0.579	-0.510	-0.439	-0.366	k_{12}
	k_{13}	-0.698	-0.588	-0.473	-0.354	-0.230	-0.653	-0.538	-0.420	-0.298	-0.173	k_{13}
0.25	k_{11}	0.070	0.023	-0.026	-0.078	-0.131	0.037	-0.013	-0.065	-0.119	-0.175	k_{11} 0.25
	k_{12}	-0.640	-0.571	-0.499	-0.425	-0.349	-0.627	-0.557	-0.485	-0.411	-0.335	k_{12}
	k_{13}	-0.710	-0.594	-0.473	-0.348	-0.218	-0.665	-0.544	-0.420	-0.292	-0.161	k_{13}
0.30	k_{11}	0.101	0.051	-0.001	-0.056	-0.112	0.068	0.015	-0.040	-0.097	-0.156	k_{11} 0.30
	k_{12}	-0.621	-0.549	-0.474	-0.397	-0.318	-0.608	-0.535	-0.460	-0.383	-0.304	k_{12}
	k_{13}	-0.722	-0.600	-0.473	-0.342	-0.206	-0.677	-0.550	-0.420	-0.286	-0.149	k_{13}
0.35	k_{11}	0.132	0.079	0.024	-0.034	-0.093	0.099	0.043	-0.015	-0.075	-0.137	k_{11} 0.35
	k_{12}	-0.602	-0.527	-0.449	-0.369	-0.287	-0.589	-0.513	-0.435	-0.355	-0.273	k_{12}
	k_{13}	-0.734	-0.606	-0.473	-0.336	-0.194	-0.689	-0.556	-0.420	-0.280	-0.137	k_{13}
0.40	k_{11}	0.163	0.107	0.049	-0.012	-0.074	0.130	0.071	0.010	-0.053	-0.118	k_{11} 0.40
	k_{12}	-0.583	-0.505	-0.424	-0.341	-0.256	-0.570	-0.491	-0.410	-0.327	-0.242	k_{12}
	k_{13}	-0.746	-0.612	-0.473	-0.330	-0.182	-0.701	-0.562	-0.420	-0.274	-0.125	k_{13}
0.45	k_{11}	0.194	0.135	0.074	0.010	-0.055	0.161	0.099	0.035	-0.031	-0.099	k_{11} 0.45
	k_{12}	-0.564	-0.483	-0.399	-0.313	-0.225	-0.551	-0.469	-0.385	-0.299	-0.211	k_{12}
	k_{13}	-0.758	-0.618	-0.473	-0.324	-0.170	-0.713	-0.568	-0.420	-0.268	-0.113	k_{13}
0.50	k_{11}	0.225	0.163	0.099	0.033	-0.036	0.192	0.127	0.060	-0.009	-0.080	k_{11} 0.50
	k_{12}	-0.545	-0.461	-0.374	-0.285	-0.194	-0.532	-0.447	-0.360	-0.271	-0.180	k_{12}
	k_{13}	-0.770	-0.624	-0.473	-0.318	-0.158	-0.725	-0.574	-0.420	-0.262	-0.101	k_{13}
0.55	k_{11}	0.256	0.191	0.124	0.055	-0.017	0.223	0.155	0.085	0.013	-0.061	k_{11} 0.55
	k_{12}	-0.526	-0.439	-0.349	-0.257	-0.163	-0.513	-0.425	-0.335	-0.243	-0.149	k_{12}
	k_{13}	-0.782	-0.630	-0.473	-0.312	-0.146	-0.737	-0.580	-0.420	-0.256	-0.089	k_{13}
0.60	k_{11}	0.287	0.219	0.149	0.077	0.002	0.254	0.183	0.110	0.035	-0.042	k_{11} 0.60
	k_{12}	-0.507	-0.417	-0.324	-0.229	-0.132	-0.494	-0.403	-0.310	-0.215	-0.118	k_{12}
	k_{13}	-0.794	-0.636	-0.473	-0.306	-0.134	-0.749	-0.586	-0.420	-0.250	-0.077	k_{13}
0.65	k_{11}	0.318	0.247	0.174	0.099	0.021	0.285	0.211	0.135	0.057	-0.023	k_{11} 0.65
	k_{12}	-0.488	-0.395	-0.299	-0.201	-0.101	-0.475	-0.381	-0.285	-0.187	-0.087	k_{12}
	k_{13}	-0.806	-0.642	-0.473	-0.300	-0.122	-0.761	-0.592	-0.420	-0.244	-0.065	k_{13}

To be continued Table V-5a on p. 102

Table V-5b   – 102 –

Table V-5a (Continuation)

m		r = 0.55					r = 0.60					m
		n = 0.40	n = 0.45	n = 0.50	n = 0.55	n = 0.60	n = 0.40	n = 0.45	n = 0.50	n = 0.55	n = 0.60	
0.70	k_{11}	0.349	0.275	0.199	0.121	0.040	0.316	0.239	0.160	0.079	− 0.004	k_{11}
	k_{12}	− 0.469	− 0.373	− 0.274	− 0.173	− 0.070	− 0.456	− 0.359	− 0.260	− 0.159	− 0.056	k_{12} 0.70
	k_{13}	− 0.818	− 0.648	− 0.473	− 0.294	− 0.110	− 0.773	− 0.598	− 0.420	− 0.238	− 0.053	k_{13}
0.75	k_{11}	0.380	0.303	0.224	0.143	0.059	0.347	0.267	0.185	0.101	0.015	k_{11}
	k_{12}	− 0.450	− 0.351	− 0.249	− 0.145	− 0.039	− 0.437	− 0.337	− 0.235	− 0.131	− 0.025	k_{12} 0.75
	k_{13}	− 0.830	− 0.654	− 0.473	− 0.288	− 0.098	− 0.785	− 0.604	− 0.420	− 0.232	− 0.041	k_{13}
0.80	k_{11}	0.411	0.331	0.249	0.165	0.078	0.378	0.295	0.210	0.123	0.034	k_{11}
	k_{12}	− 0.431	− 0.329	− 0.224	− 0.117	− 0.008	− 0.418	− 0.315	− 0.210	− 0.103	0.006	k_{12} 0.80
	k_{13}	− 0.842	− 0.660	− 0.473	− 0.282	− 0.086	− 0.797	− 0.610	− 0.420	− 0.226	− 0.029	k_{13}
0.85	k_{11}	0.442	0.359	0.274	0.187	0.097	0.409	0.323	0.235	0.145	0.053	k_{11}
	k_{12}	− 0.412	− 0.307	− 0.199	− 0.089	0.023	− 0.399	− 0.293	− 0.185	− 0.075	0.037	k_{12} 0.85
	k_{13}	− 0.854	− 0.666	− 0.473	− 0.276	− 0.074	− 0.809	− 0.616	− 0.420	− 0.220	− 0.017	k_{13}
0.90	k_{11}	0.473	0.387	0.299	0.209	0.116	0.440	0.351	0.260	0.167	0.072	k_{11}
	k_{12}	− 0.393	− 0.285	− 0.174	− 0.061	0.054	− 0.380	− 0.271	− 0.160	− 0.047	0.068	k_{12} 0.90
	k_{13}	− 0.866	− 0.672	− 0.473	− 0.270	− 0.062	− 0.821	− 0.622	− 0.420	− 0.214	− 0.005	k_{13}
0.95	k_{11}	0.504	0.415	0.324	0.231	0.135	0.471	0.379	0.285	0.189	0.091	k_{11}
	k_{12}	− 0.374	− 0.263	− 0.149	− 0.033	0.085	− 0.361	− 0.249	− 0.135	− 0.019	0.099	k_{12} 0.95
	k_{13}	− 0.878	− 0.678	− 0.473	− 0.264	− 0.050	− 0.833	− 0.628	− 0.420	− 0.208	0.007	k_{13}
1.00	k_{11}	0.535	0.443	0.349	0.252	0.154	0.502	0.407	0.310	0.211	0.110	k_{11}
	k_{12}	− 0.355	− 0.241	− 0.124	− 0.005	0.116	− 0.342	− 0.227	− 0.110	0.009	0.130	k_{12} 1.00
	k_{13}	− 0.890	− 0.684	− 0.473	− 0.258	− 0.038	− 0.845	− 0.634	− 0.420	− 0.202	0.019	k_{13}

Table V-5b

Top number = k_{11};   Bottom number = k_{12};   Medium number = k_{13}

m		r = 0.65					r = 0.70					m
		n = 0.40	n = 0.45	n = 0.50	n = 0.55	n = 0.60	n = 0.40	n = 0.45	n = 0.50	n = 0.55	n = 0.60	
−0.10	k_{11}	− 0.212	− 0.245	− 0.279	− 0.314	− 0.351	− 0.245	− 0.281	− 0.318	− 0.356	− 0.395	k_{11}
	k_{12}	− 0.748	− 0.698	− 0.646	− 0.593	− 0.539	− 0.735	− 0.684	− 0.633	− 0.579	− 0.525	k_{12} −0.10
	k_{13}	− 0.535	− 0.453	− 0.368	− 0.279	− 0.187	− 0.490	− 0.404	− 0.315	− 0.224	− 0.130	k_{13}
−0.05	k_{11}	− 0.181	− 0.217	− 0.254	− 0.292	− 0.332	− 0.214	− 0.253	− 0.293	− 0.334	− 0.376	k_{11}
	k_{12}	− 0.729	− 0.676	− 0.621	− 0.565	− 0.508	− 0.716	− 0.662	− 0.608	− 0.551	− 0.494	k_{12} −0.05
	k_{13}	− 0.547	− 0.459	− 0.368	− 0.273	− 0.175	− 0.502	− 0.410	− 0.315	− 0.218	− 0.118	k_{13}
0	k_{11}	− 0.150	− 0.189	− 0.229	− 0.270	− 0.313	− 0.183	− 0.225	− 0.268	− 0.312	− 0.357	k_{11}
	k_{12}	− 0.710	− 0.654	− 0.596	− 0.537	− 0.477	− 0.697	− 0.640	− 0.583	− 0.523	− 0.463	k_{12} 0
	k_{13}	− 0.559	− 0.465	− 0.368	− 0.267	− 0.163	− 0.514	− 0.416	− 0.315	− 0.212	− 0.106	k_{13}
0.05	k_{11}	− 0.119	− 0.161	− 0.204	− 0.248	− 0.294	− 0.152	− 0.197	− 0.243	− 0.290	− 0.338	k_{11}
	k_{12}	− 0.691	− 0.632	− 0.571	− 0.509	− 0.446	− 0.678	− 0.618	− 0.558	− 0.495	−- 0.432	k_{12} 0.05
	k_{13}	− 0.571	− 0.471	− 0.368	− 0.261	− 0.151	− 0.526	− 0.422	− 0.315	− 0.206	− 0.094	k_{13}
0.10	k_{11}	− 0.088	− 0.133	− 0.179	− 0.226	− 0.275	− 0.121	− 0.169	− 0.218	− 0.268	− 0.319	k_{11}
	k_{12}	− 0.672	− 0.610	− 0.546	− 0.481	− 0.415	− 0.659	− 0.596	− 0.533	− 0.467	− 0.401	k_{12} 0.10
	k_{13}	− 0.583	− 0.477	− 0.368	− 0.255	− 0.139	− 0.538	− 0.428	− 0.315	− 0.200	− 0.082	k_{13}
0.15	k_{11}	− 0.057	− 0.105	− 0.154	− 0.204	− 0.256	− 0.090	− 0.141	− 0.193	− 0.246	− 0.300	k_{11}
	k_{12}	− 0.653	− 0.588	− 0.521	− 0.453	− 0.384	− 0.640	− 0.574	− 0.508	− 0.439	− 0.370	k_{12} 0.15
	k_{13}	− 0.595	− 0.483	− 0.368	− 0.249	− 0.127	− 0.550	− 0.434	− 0.315	− 0.194	− 0.070	k_{13}
0.20	k_{11}	− 0.026	− 0.077	− 0.129	− 0.182	− 0.237	− 0.059	− 0.113	− 0.168	− 0.224	− 0.281	k_{11}
	k_{12}	− 0.634	− 0.566	− 0.496	− 0.425	− 0.353	− 0.621	− 0.552	− 0.483	− 0.411	− 0.339	k_{12} 0.20
	k_{13}	− 0.607	− 0.489	− 0.368	− 0.243	− 0.115	− 0.562	− 0.440	− 0.315	− 0.188	− 0.058	k_{13}

Tabelle V-5b (Continuation)

m	$r = 0.65$					$r = 0.70$					m
	$n=0.40$	$n=0.45$	$n=0.50$	$n=0.55$	$n=0.60$	$n=0.40$	$n=0.45$	$n=0.50$	$n=0.55$	$n=0.60$	
0.25 k_{11} k_{12} k_{13}	0.005 -0.615 -0.619	-0.049 -0.544 -0.495	-0.104 -0.471 -0.368	-0.160 -0.397 -0.237	-0.218 -0.322 -0.103	-0.028 -0.602 -0.574	-0.085 -0.530 -0.446	-0.143 -0.458 -0.315	-0.202 -0.383 -0.182	-0.262 -0.308 -0.046	k_{11} k_{12} k_{13} 0.25
0.30 k_{11} k_{12} k_{13}	0.036 -0.596 -0.631	-0.021 -0.522 -0.501	-0.079 -0.446 -0.368	-0.138 -0.369 -0.231	-0.199 -0.291 -0.091	0.003 -0.583 -0.586	-0.057 -0.508 -0.452	-0.118 -0.433 -0.315	-0.180 -0.355 -0.176	-0.243 -0.277 -0.034	k_{11} k_{12} k_{13} 0.30
0.35 k_{11} k_{12} k_{13}	0.067 -0.577 -0.643	0.007 -0.500 -0.507	-0.054 -0.421 -0.368	-0.116 -0.341 -0.225	-0.180 -0.260 -0.080	0.034 -0.564 -0.598	-0.029 -0.486 -0.458	-0.093 -0.408 -0.315	-0.158 -0.327 -0.170	-0.224 -0.246 -0.022	k_{11} k_{12} k_{13} 0.35
0.40 k_{11} k_{12} k_{13}	0.098 -0.558 -0.655	0.035 -0.478 -0.513	-0.029 -0.396 -0.368	-0.094 -0.313 -0.219	-0.161 -0.229 -0.067	0.065 -0.545 -0.610	-0.001 -0.464 -0.464	-0.068 -0.383 -0.315	-0.136 -0.299 -0.164	-0.205 -0.215 -0.010	k_{11} k_{12} k_{13} 0.40
0.45 k_{11} k_{12} k_{13}	0.129 -0.539 -0.667	0.063 -0.456 -0.519	-0.004 -0.371 -0.368	-0.072 -0.285 -0.213	-0.142 -0.198 -0.055	0.096 -0.526 -0.622	0.027 -0.442 -0.470	-0.043 -0.358 -0.315	-0.114 -0.271 -0.158	-0.186 -0.184 0.002	k_{11} k_{12} k_{13} 0.45
0.50 k_{11} k_{12} k_{13}	0.160 -0.520 -0.679	0.091 -0.434 -0.525	0.021 -0.346 -0.368	-0.050 -0.257 -0.207	-0.123 -0.167 -0.043	0.127 -0.507 -0.634	0.055 -0.420 -0.476	-0.018 -0.333 -0.315	-0.092 -0.243 -0.152	-0.167 -0.153 0.014	k_{11} k_{12} k_{13} 0.50
0.55 k_{11} k_{12} k_{13}	0.191 -0.501 -0.691	0.119 -0.412 -0.531	0.046 -0.321 -0.368	-0.028 -0.229 -0.201	-0.104 -0.136 -0.031	0.158 -0.488 -0.646	0.083 -0.398 -0.482	0.008 -0.308 -0.315	-0.070 -0.215 -0.146	-0.148 -0.122 0.026	k_{11} k_{12} k_{13} 0.55
0.60 k_{11} k_{12} k_{13}	0.222 -0.482 -0.703	0.147 -0.390 -0.537	0.071 -0.296 -0.368	-0.006 -0.201 -0.195	-0.085 -0.105 -0.019	0.189 -0.469 -0.658	0.111 -0.376 -0.488	0.033 -0.283 -0.315	-0.048 -0.187 -0.140	-0.129 -0.091 0.038	k_{11} k_{12} k_{13} 0.60
0.65 k_{11} k_{12} k_{13}	0.253 -0.463 -0.715	0.175 -0.368 -0.543	0.096 -0.271 -0.368	0.016 -0.173 -0.189	-0.066 -0.074 -0.007	0.220 -0.449 -0.670	0.139 -0.354 -0.494	0.058 -0.258 -0.315	-0.026 -0.159 -0.134	-0.110 -0.060 0.050	k_{11} k_{12} k_{13} 0.65
0.70 k_{11} k_{12} k_{13}	0.284 -0.444 -0.727	0.203 -0.346 -0.549	0.121 -0.246 -0.368	0.038 -0.145 -0.183	-0.047 -0.043 0.005	0.251 -0.431 -0.682	0.167 -0.332 -0.500	0.083 -0.233 -0.315	-0.004 -0.131 -0.128	-0.091 -0.029 0.062	k_{11} k_{12} k_{13} 0.70
0.75 k_{11} k_{12} k_{13}	0.315 -0.425 -0.739	0.231 -0.324 -0.555	0.146 -0.221 -0.368	0.060 -0.117 -0.177	-0.028 -0.012 0.017	0.282 -0.412 -0.694	0.195 -0.310 -0.506	0.108 -0.208 -0.315	0.018 -0.103 -0.122	-0.072 0.002 0.074	k_{11} k_{12} k_{13} 0.75
0.80 k_{11} k_{12} k_{13}	0.346 -0.406 -0.751	0.259 -0.302 -0.561	0.171 -0.196 -0.368	0.082 -0.089 -0.171	-0.009 0.019 0.029	0.313 -0.393 -0.706	0.223 -0.288 -0.512	0.133 -0.183 -0.315	0.040 -0.075 -0.116	-0.053 0.033 0.086	k_{11} k_{12} k_{13} 0.80
0.85 k_{11} k_{12} k_{13}	0.377 -0.387 -0.763	0.287 -0.280 -0.567	0.196 -0.171 -0.368	0.104 -0.061 -0.165	0.010 0.050 0.041	0.344 -0.374 -0.718	0.251 -0.266 -0.518	0.158 -0.158 -0.315	0.062 -0.047 -0.110	-0.034 0.064 0.098	k_{11} k_{12} k_{13} 0.85
0.90 k_{11} k_{12} k_{13}	0.408 -0.368 -0.775	0.315 -0.258 -0.573	0.221 -0.146 -0.368	0.126 -0.033 -0.159	0.029 0.081 0.053	0.375 -0.355 -0.730	0.279 -0.244 -0.524	0.183 -0.133 -0.315	0.084 -0.019 -0.104	-0.015 0.095 0.110	k_{11} k_{12} k_{13} 0.90
0.95 k_{11} k_{12} k_{13}	0.439 -0.349 -0.787	0.343 -0.236 -0.579	0.246 -0.121 -0.368	0.148 -0.005 -0.153	0.048 0.112 0.065	0.406 -0.336 -0.742	0.307 -0.222 -0.530	0.208 -0.108 -0.315	0.106 0.009 -0.098	0.004 0.126 0.122	k_{11} k_{12} k_{13} 0.95
1.00 k_{11} k_{12} k_{13}	0.470 -0.330 -0.799	0.371 -0.214 -0.585	0.271 -0.096 -0.368	0.170 0.023 -0.147	0.067 0.143 0.077	0.437 -0.317 -0.754	0.335 -0.200 -0.536	0.233 -0.083 -0.315	0.128 0.037 -0.092	0.023 0.157 0.134	k_{11} k_{12} k_{13} 1.00

Table V-5c — 104 —

Table V-5c

Top number $= k_{11}$; Medium number $= k_{12}$; Bottom number $= k_{13}$

m		$r = 0.75$					$r = 0.80$					m	
		$n=0.40$	$n=0.45$	$n=0.50$	$n=0.55$	$n=0.60$	$n=0.40$	$n=0.45$	$n=0.50$	$n=0.55$	$n=0.60$		
-0.10	k_{11}	-0.278	-0.317	-0.356	-0.397	-0.439	-0.311	-0.353	-0.395	-0.439	-0.483	k_{11}	-0.10
	k_{12}	-0.722	-0.671	-0.619	-0.565	-0.511	-0.709	-0.658	-0.605	-0.552	-0.497	k_{12}	
	k_{13}	-0.444	-0.354	-0.263	-0.168	-0.072	-0.398	-0.305	-0.210	-0.113	-0.014	k_{13}	
-0.05	k_{11}	-0.247	-0.289	-0.331	-0.375	-0.420	-0.280	-0.325	-0.370	-0.417	-0.464	k_{11}	-0.05
	k_{12}	-0.703	-0.649	-0.594	-0.537	-0.480	-0.690	-0.636	-0.580	-0.524	-0.466	k_{12}	
	k_{13}	-0.456	-0.360	-0.263	-0.162	-0.060	-0.410	-0.311	-0.210	-0.107	-0.002	k_{13}	
0	k_{11}	-0.216	-0.261	-0.306	-0.353	-0.401	-0.249	-0.297	-0.345	-0.395	-0.445	k_{11}	0
	k_{12}	-0.684	-0.627	-0.569	-0.509	-0.449	-0.671	-0.614	-0.555	-0.496	-0.435	k_{12}	
	k_{13}	-0.468	-0.366	-0.263	-0.156	-0.048	-0.422	-0.317	-0.210	-0.101	0.010	k_{13}	
0.05	k_{11}	-0.185	-0.233	-0.281	-0.331	-0.382	-0.218	-0.269	-0.320	-0.373	-0.426	k_{11}	0.05
	k_{12}	-0.665	-0.605	-0.544	-0.481	-0.418	-0.652	-0.592	-0.530	-0.468	-0.404	k_{12}	
	k_{13}	-0.480	-0.372	-0.263	-0.150	-0.036	-0.434	-0.323	-0.210	-0.095	0.022	k_{13}	
0.10	k_{11}	-0.154	-0.205	-0.256	-0.309	-0.363	-0.187	-0.241	-0.295	-0.351	-0.407	k_{11}	0.10
	k_{12}	-0.646	-0.583	-0.519	-0.453	-0.387	-0.633	-0.570	-0.505	-0.440	-0.373	k_{12}	
	k_{13}	-0.492	-0.378	-0.263	-0.144	-0.024	-0.446	-0.329	-0.210	-0.089	0.034	k_{13}	
0.15	k_{11}	-0.123	-0.177	-0.231	-0.287	-0.344	-0.156	-0.213	-0.270	-0.329	-0.388	k_{11}	0.15
	k_{12}	-0.627	-0.561	-0.494	-0.425	-0.356	-0.614	-0.548	-0.480	-0.412	-0.342	k_{12}	
	k_{13}	-0.504	-0.384	-0.263	-0.138	-0.012	-0.458	-0.335	-0.210	-0.083	0.046	k_{13}	
0.20	k_{11}	-0.092	-0.149	-0.206	-0.265	-0.325	-0.125	-0.185	-0.245	-0.307	-0.369	k_{11}	0.20
	k_{12}	-0.608	-0.539	-0.469	-0.397	-0.325	-0.595	-0.526	-0.455	-0.384	-0.311	k_{12}	
	k_{13}	-0.516	-0.390	-0.263	-0.132	0.000	-0.470	-0.341	-0.210	-0.077	0.058	k_{13}	
0.25	k_{11}	-0.061	-0.121	-0.181	-0.243	-0.306	-0.094	-0.157	-0.220	-0.285	-0.350	k_{11}	0.25
	k_{12}	-0.589	-0.517	-0.444	-0.369	-0.294	-0.576	-0.504	-0.430	-0.356	-0.280	k_{12}	
	k_{13}	-0.528	-0.396	-0.263	-0.126	0.012	-0.482	-0.347	-0.210	-0.071	0.070	k_{13}	
0.30	k_{11}	-0.030	-0.093	-0.156	-0.221	-0.287	-0.063	-0.129	-0.195	-0.262	-0.331	k_{11}	0.30
	k_{12}	-0.570	-0.495	-0.419	-0.341	-0.263	-0.557	-0.482	-0.405	-0.328	-0.249	k_{12}	
	k_{13}	-0.540	-0.402	-0.263	-0.120	0.024	-0.494	-0.353	-0.210	-0.065	0.082	k_{13}	
0.35	k_{11}	0.001	-0.065	-0.131	-0.199	-0.268	-0.032	-0.101	-0.170	-0.241	-0.312	k_{11}	0.35
	k_{12}	-0.551	-0.473	-0.394	-0.313	-0.232	-0.538	-0.460	-0.380	-0.300	-0.218	k_{12}	
	k_{13}	-0.552	-0.408	-0.263	-0.114	0.036	-0.506	-0.359	-0.210	-0.059	0.094	k_{13}	
0.40	k_{11}	0.032	-0.037	-0.106	-0.177	-0.249	-0.001	-0.073	-0.145	-0.219	-0.293	k_{11}	0.40
	k_{12}	-0.532	-0.451	-0.369	-0.285	-0.201	-0.519	-0.438	-0.355	-0.272	-0.187	k_{13}	
	k_{13}	-0.564	-0.414	-0.263	-0.108	0.048	-0.518	-0.365	-0.210	-0.053	0.106	k_{13}	
0.45	k_{11}	0.063	-0.009	-0.081	-0.155	-0.230	0.030	-0.045	-0.120	-0.197	-0.274	k_{11}	0.45
	k_{12}	-0.513	-0.429	-0.344	-0.257	-0.170	-0.500	-0.416	-0.330	-0.244	-0.156	k_{12}	
	k_{13}	-0.576	-0.420	-0.263	-0.102	0.060	-0.530	-0.371	-0.210	-0.047	0.118	k_{13}	
0.50	k_{11}	0.094	0.019	-0.056	-0.133	-0.211	0.061	-0.017	-0.095	-0.175	-0.255	k_{11}	0.50
	k_{12}	-0.494	-0.407	-0.319	-0.229	-0.139	-0.481	-0.394	-0.305	-0.216	-0.125	k_{12}	
	k_{13}	-0.588	-0.426	-0.263	-0.096	0.072	-0.542	-0.377	-0.210	-0.041	0.130	k_{13}	
0.55	k_{11}	0.125	0.047	-0.031	-0.111	-0.192	0.092	0.012	-0.070	-0.153	-0.236	k_{11}	0.55
	k_{12}	-0.475	-0.385	-0.294	-0.201	-0.108	-0.462	-0.372	-0.280	-0.188	-0.094	k_{12}	
	k_{13}	-0.600	-0.432	-0.263	-0.090	0.084	-0.554	-0.383	-0.210	-0.035	0.142	k_{13}	
0.60	k_{11}	0.156	0.075	-0.006	-0.089	-0.173	0.123	0.040	-0.045	-0.131	-0.217	k_{11}	0.60
	k_{12}	-0.456	-0.363	-0.269	-0.173	-0.077	-0.443	-0.350	-0.255	-0.160	-0.063	k_{12}	
	k_{13}	-0.612	-0.438	-0.263	-0.084	0.096	-0.566	-0.389	-0.210	-0.029	0.154	k_{13}	
0.65	k_{11}	0.187	0.103	0.019	-0.067	-0.154	0.154	0.068	-0.020	-0.109	-0.198	k_{11}	0.65
	k_{12}	-0.437	-0.341	-0.244	-0.145	-0.046	-0.424	-0.328	-0.230	-0.132	-0.032	k_{12}	
	k_{13}	-0.624	-0.444	-0.263	-0.078	0.108	-0.578	-0.395	-0.210	-0.023	0.166	k_{13}	

Table V-5c (Continuation)

m		r = 0.75					r = 0.80					m
		n = 0.40	n = 0.45	n = 0.50	n = 0.55	n = 0.60	n = 0.40	n = 0.45	n = 0.50	n = 0.55	n = 0.60	
0.70	k_{11}	0.218	0.131	0.044	−0.045	−0.135	0.185	0.096	0.005	−0.087	−0.179	k_{11} 0.70
	k_{12}	−0.418	−0.319	−0.219	−0.117	−0.015	−0.405	−0.306	−0.205	−0.104	−0.001	k_{12}
	k_{13}	−0.636	−0.450	−0.263	−0.072	0.120	−0.590	−0.401	−0.210	−0.017	0.178	k_{13}
0.75	k_{11}	0.249	0.159	0.069	−0.023	−0.116	0.216	0.124	0.030	−0.065	−0.160	k_{11} 0.75
	k_{12}	−0.399	−0.297	−0.194	−0.089	0.016	−0.386	−0.284	−0.180	−0.076	0.030	k_{12}
	k_{13}	−0.648	−0.456	−0.263	−0.066	0.132	−0.602	−0.407	−0.210	−0.011	0.190	k_{13}
0.80	k_{11}	0.280	0.187	0.094	−0.001	−0.097	0.247	0.152	0.055	−0.043	−0.141	k_{11} 0.80
	k_{12}	−0.380	−0.275	−0.169	−0.061	0.047	−0.367	−0.262	−0.155	−0.048	0.061	k_{12}
	k_{13}	−0.660	−0.462	−0.263	−0.060	0.144	−0.614	−0.413	−0.210	−0.005	0.202	k_{13}
0.85	k_{11}	0.311	0.215	0.119	0.021	−0.078	0.278	0.180	0.080	−0.021	−0.122	k_{11} 0.85
	k_{12}	−0.361	−0.253	−0.144	−0.033	0.078	−0.348	−0.240	−0.130	−0.020	0.092	k_{12}
	k_{13}	−0.672	−0.468	−0.263	−0.054	0.156	−0.626	−0.419	−0.210	0.001	0.214	k_{13}
0.90	k_{11}	0.342	0.243	0.144	0.043	−0.059	0.309	0.208	0.105	0.002	−0.103	k_{11} 0.90
	k_{12}	−0.342	−0.231	−0.119	−0.005	0.109	−0.329	−0.218	−0.105	0.008	0.123	k_{12}
	k_{13}	−0.684	−0.474	−0.263	−0.048	0.168	−0.638	−0.425	−0.210	0.007	0.226	k_{13}
0.95	k_{11}	0.373	0.271	0.169	0.065	−0.040	0.340	0.236	0.130	0.024	−0.084	k_{11} 0.95
	k_{12}	−0.323	−0.209	−0.094	0.023	0.140	−0.310	−0.196	−0.080	0.036	0.154	k_{12}
	k_{13}	−0.696	−0.480	−0.263	−0.042	0.180	−0.650	−0.431	−0.210	0.013	0.238	k_{13}
1.00	k_{11}	0.404	0.299	0.194	0.087	−0.021	0.371	0.264	0.155	0.046	−0.065	k_{11} 1.00
	k_{12}	−0.304	−0.187	−0.069	0.051	0.171	−0.291	−0.174	−0.055	0.064	0.185	k_{12}
	k_{13}	−0.708	−0.486	−0.263	−0.036	0.192	−0.662	−0.437	−0.210	0.019	0.250	k_{13}

Table V-5d

Top number = k_{11};   Medium number = k_{12};   Bottom number = k_{13}

m		r = 0.85					r = 0.90					m
		n = 0.40	n = 0.45	n = 0.50	n = 0.55	n = 0.60	n = 0.40	n = 0.45	n = 0.50	n = 0.55	n = 0.60	
−0.10	k_{11}	−0.344	−0.388	−0.434	−0.480	−0.527	−0.376	−0.424	−0.473	−0.521	−0.570	k_{11} −0.10
	k_{12}	−0.696	−0.644	−0.591	−0.538	−0.483	−0.684	−0.631	−0.578	−0.524	−0.470	k_{12}
	k_{13}	−0.353	−0.256	−0.158	−0.058	0.043	−0.307	−0.207	−0.105	−0.003	0.101	k_{13}
−0.05	k_{11}	−0.313	−0.360	−0.409	−0.458	−0.508	−0.345	−0.396	−0.448	−0.499	−0.551	k_{11} −0.05
	k_{12}	−0.677	−0.622	−0.566	−0.510	−0.452	−0.665	−0.609	−0.553	−0.496	−0.439	k_{12}
	k_{13}	−0.365	−0.262	−0.158	−0.052	0.055	−0.319	−0.212	−0.105	0.003	0.113	k_{13}
0	k_{11}	−0.282	−0.332	−0.384	−0.436	−0.489	−0.314	−0.368	−0.423	−0.477	−0.532	k_{11} 0
	k_{12}	−0.658	−0.600	−0.541	−0.482	−0.421	−0.646	−0.587	−0.528	−0.468	−0.408	k_{12}
	k_{13}	−0.377	−0.268	−0.158	−0.046	0.067	−0.331	−0.219	−0.105	0.009	0.125	k_{13}
0.05	k_{11}	−0.251	−0.304	−0.359	−0.414	−0.470	−0.283	−0.340	−0.398	−0.455	−0.513	k_{11} 0.05
	k_{12}	−0.639	−0.578	−0.516	−0.454	−0.390	−0.627	−0.565	−0.503	−0.440	−0.377	k_{12}
	k_{13}	−0.389	−0.274	−0.158	−0.040	0.079	−0.343	−0.225	−0.105	0.015	0.137	k_{13}
0.10	k_{11}	−0.220	−0.276	−0.334	−0.392	−0.451	−0.252	−0.312	−0.373	−0.433	−0.494	k_{11} 0.10
	k_{12}	−0.620	−0.556	−0.491	−0.426	−0.359	−0.608	−0.543	−0.478	−0.412	−0.346	k_{12}
	k_{13}	−0.401	−0.280	−0.158	−0.034	0.091	−0.355	−0.231	−0.105	0.021	0.149	k_{13}
0.15	k_{11}	−0.189	−0.248	−0.309	−0.370	−0.432	−0.221	−0.284	−0.348	−0.411	−0.475	k_{11} 0.15
	k_{12}	−0.601	−0.534	−0.466	−0.398	−0.328	−0.589	−0.521	−0.453	−0.384	−0.315	k_{12}
	k_{13}	−0.413	−0.286	−0.158	−0.028	0.103	−0.367	−0.237	−0.105	0.027	0.161	k_{13}
0.20	k_{11}	−0.158	−0.220	−0.284	−0.348	−0.413	−0.190	−0.256	−0.323	−0.389	−0.456	k_{11} 0.20
	k_{12}	−0.582	−0.512	−0.441	−0.370	−0.297	−0.570	−0.499	−0.428	−0.356	−0.284	k_{12}
	k_{13}	−0.425	−0.292	−0.158	−0.022	0.115	−0.379	−0.242	−0.105	0.033	0.173	k_{13}

To be continued Table V-5d on p. 106

Table V-5d — 106 —

Table V-5d (Continuation)

m		$r = 0.85$					$r = 0.90$					m
		$n = 0.40$	$n = 0.45$	$n = 0.50$	$n = 0.55$	$n = 0.60$	$n = 0.40$	$n = 0.45$	$n = 0.50$	$n = 0.55$	$n = 0.60$	
0.25	k_{11}	-0.127	-0.192	-0.259	-0.326	-0.394	-0.159	-0.228	-0.298	-0.367	-0.437	k_{11} 0.25
	k_{12}	-0.563	-0.490	-0.416	-0.342	-0.266	-0.551	-0.477	-0.403	-0.328	-0.253	k_{12}
	k_{13}	-0.437	-0.298	-0.158	-0.016	0.127	-0.391	-0.249	-0.105	0.039	0.185	k_{13}
0.30	k_{11}	-0.096	-0.164	-0.234	-0.304	-0.375	-0.128	-0.200	-0.273	-0.345	-0.418	k_{11} 0.30
	k_{12}	-0.544	-0.468	-0.391	-0.314	-0.235	-0.532	-0.455	-0.378	-0.300	-0.222	k_{12}
	k_{13}	-0.449	-0.304	-0.158	-0.010	0.139	-0.403	-0.255	-0.105	0.045	0.197	k_{13}
0.35	k_{11}	-0.065	-0.136	-0.209	-0.282	-0.356	-0.097	-0.172	-0.248	-0.323	-0.399	k_{11} 0.35
	k_{12}	-0.525	-0.446	-0.366	-0.286	-0.204	-0.513	-0.433	-0.353	-0.272	-0.191	k_{12}
	k_{13}	-0.461	-0.310	-0.158	-0.004	0.151	-0.415	-0.261	-0.105	0.051	0.209	k_{13}
0.40	k_{11}	-0.034	-0.108	-0.184	-0.260	-0.337	-0.066	-0.144	-0.223	-0.301	-0.380	k_{11} 0.40
	k_{12}	-0.506	-0.424	-0.341	-0.258	-0.173	-0.494	-0.411	-0.328	-0.244	-0.160	k_{12}
	k_{13}	-0.473	-0.316	-0.158	0.002	0.163	-0.427	-0.267	-0.105	0.057	0.221	k_{13}
0.45	k_{11}	-0.003	-0.080	-0.159	-0.238	-0.318	-0.035	-0.116	-0.198	-0.279	-0.361	k_{11} 0.45
	k_{12}	-0.487	-0.402	-0.316	-0.230	-0.142	-0.475	-0.389	-0.303	-0.216	-0.129	k_{12}
	k_{13}	-0.485	-0.322	-0.158	0.008	0.175	-0.439	-0.273	-0.105	0.063	0.233	k_{13}
0.50	k_{11}	0.028	-0.052	-0.134	-0.216	-0.299	-0.004	-0.088	-0.173	-0.257	-0.342	k_{11} 0.50
	k_{12}	-0.468	-0.380	-0.291	-0.202	-0.111	-0.456	-0.367	-0.278	-0.188	-0.098	k_{12}
	k_{13}	-0.497	-0.328	-0.158	0.014	0.187	-0.451	-0.279	-0.105	0.069	0.245	k_{13}
0.55	k_{11}	0.059	-0.024	-0.109	-0.194	-0.280	0.027	-0.060	-0.148	-0.235	-0.323	k_{11} 0.55
	k_{12}	-0.449	-0.358	-0.266	-0.174	-0.080	-0.437	-0.345	-0.253	-0.160	-0.067	k_{12}
	k_{13}	-0.509	-0.334	-0.158	0.020	0.199	-0.463	-0.285	-0.105	0.075	0.257	k_{13}
0.60	k_{11}	0.090	0.004	-0.084	-0.172	-0.261	0.058	-0.032	-0.123	-0.213	-0.304	k_{11} 0.60
	k_{12}	-0.430	-0.336	-0.241	-0.146	-0.049	-0.418	-0.323	-0.228	-0.132	-0.036	k_{12}
	k_{13}	-0.521	-0.340	-0.158	0.026	0.211	-0.475	-0.291	-0.105	0.081	0.269	k_{13}
0.65	k_{11}	0.121	0.032	-0.059	-0.150	-0.242	0.089	-0.004	-0.098	-0.191	-0.285	k_{11} 0.65
	k_{12}	-0.411	-0.314	-0.212	-0.118	-0.018	-0.399	-0.301	-0.203	-0.104	-0.005	k_{12}
	k_{13}	-0.533	-0.346	-0.158	0.032	0.223	-0.487	-0.297	-0.105	0.087	0.281	k_{13}
0.70	k_{11}	0.152	0.060	-0.034	-0.128	-0.223	0.120	0.024	-0.073	-0.169	-0.266	k_{11} 0.70
	k_{12}	-0.392	-0.292	-0.191	-0.090	0.013	-0.380	-0.279	-0.178	-0.076	0.026	k_{12}
	k_{13}	-0.545	-0.352	-0.158	0.038	0.235	-0.499	-0.304	-0.105	0.093	0.293	k_{13}
0.75	k_{11}	0.183	0.088	-0.009	-0.106	-0.204	0.151	0.052	-0.048	-0.147	-0.247	k_{11} 0.75
	k_{12}	-0.373	-0.270	-0.166	-0.062	0.044	-0.361	-0.257	-0.153	-0.048	0.057	k_{12}
	k_{13}	-0.557	-0.358	-0.158	0.044	0.247	-0.511	-0.309	-0.105	0.099	0.305	k_{13}
0.80	k_{11}	0.214	0.116	0.016	-0.084	-0.185	0.182	0.080	-0.023	-0.125	-0.228	k_{11} 0.80
	k_{12}	-0.354	-0.248	-0.141	-0.034	0.075	-0.342	-0.235	-0.128	-0.020	0.088	k_{12}
	k_{13}	-0.569	-0.364	-0.158	0.050	0.259	-0.523	-0.315	-0.105	0.105	0.317	k_{13}
0.85	k_{11}	0.245	0.144	0.041	-0.062	-0.166	0.213	0.108	0.003	-0.103	-0.209	k_{11} 0.85
	k_{12}	-0.335	-0.226	-0.116	-0.006	0.106	-0.323	-0.213	-0.103	0.008	0.119	k_{12}
	k_{13}	-0.581	-0.370	-0.158	0.056	0.271	-0.535	-0.321	-0.105	0.111	0.329	k_{13}
0.90	k_{11}	0.276	0.172	0.066	-0.040	-0.147	0.244	0.136	0.028	-0.081	-0.190	k_{11} 0.90
	k_{12}	-0.316	-0.204	-0.091	0.022	0.137	-0.304	-0.191	-0.078	0.036	0.150	k_{12}
	k_{13}	-0.593	-0.376	-0.158	0.062	0.283	-0.547	-0.327	-0.105	0.117	0.341	k_{13}
0.95	k_{11}	0.307	0.200	0.091	-0.018	-0.128	0.275	0.164	0.053	-0.059	-0.171	k_{11} 0.95
	k_{12}	-0.297	-0.182	-0.066	0.050	0.168	-0.286	-0.169	-0.053	0.064	0.181	k_{12}
	k_{13}	-0.605	-0.382	-0.158	0.068	0.295	-0.559	-0.333	-0.105	0.123	0.353	k_{13}
1.00	k_{11}	0.338	0.228	0.116	0.004	-0.109	0.306	0.192	0.078	-0.037	-0.152	k_{11} 1.00
	k_{12}	-0.278	-0.160	-0.041	0.078	0.199	-0.266	-0.147	-0.028	0.092	0.212	k_{12}
	k_{13}	-0.617	-0.388	-0.158	0.074	0.307	-0.571	-0.339	-0.105	0.129	0.365	k_{13}

Table V-5e

Top number = k_{11};   Medium number = k_{12};   Bottom number = k_{13}

m		r = 0.95					r = 1.00					m	
		n = 0.40	n = 0.45	n = 0.50	n = 0.55	n = 0.60	n = 0.40	n = 0.45	n = 0.50	n = 0.55	n = 0.60		
	k_{11}	− 0.409	− 0.460	− 0.511	− 0.563	− 0.614	− 0.442	− 0.496	− 0.550	− 0.604	− 0.658	k_{11}	
− 0.10	k_{12}	− 0.671	− 0.617	− 0.564	− 0.510	− 0.456	− 0.658	− 0.604	− 0.550	− 0.496	− 0.442	k_{12}	− 0.10
	k_{13}	− 0.262	− 0.157	− 0.053	0.053	0.158	− 0.216	− 0.108	0.000	0.108	0.216	k_{13}	
	k_{11}	− 0.378	− 0.432	− 0.486	− 0.541	− 0.595	− 0.411	− 0.468	− 0.525	− 0.582	− 0.639	k_{11}	
− 0.05	k_{12}	− 0.652	− 0.595	− 0.539	− 0.482	− 0.425	− 0.639	− 0.582	− 0.525	− 0.468	− 0.411	k_{12}	− 0.05
	k_{13}	− 0.274	− 0.163	− 0.053	0.059	0.170	− 0.228	− 0.114	0.000	0.114	0.228	k_{13}	
	k_{11}	− 0.347	− 0.404	− 0.461	− 0.519	− 0.576	− 0.380	− 0.440	− 0.500	− 0.560	− 0.620	k_{11}	
0.00	k_{12}	− 0.633	− 0.573	− 0.514	− 0.454	− 0.394	− 0.620	− 0.560	− 0.500	− 0.440	− 0.380	k_{12}	0.00
	k_{13}	− 0.286	− 0.169	− 0.053	0.065	0.182	− 0.240	− 0.120	0.000	0.120	0.240	k_{13}	
	k_{11}	− 0.316	− 0.376	− 0.436	− 0.497	− 0.557	− 0.349	− 0.412	− 0.475	− 0.538	− 0.601	k_{11}	
0.05	k_{12}	− 0.614	− 0.551	− 0.489	− 0.426	− 0.363	− 0.601	− 0.538	− 0.475	− 0.412	− 0.349	k_{12}	0.05
	k_{13}	− 0.298	− 0.175	− 0.053	0.071	0.194	− 0.252	− 0.126	0.000	0.126	0.252	k_{13}	
	k_{11}	− 0.285	− 0.348	− 0.411	− 0.475	− 0.538	− 0.318	− 0.384	− 0.450	− 0.516	− 0.582	k_{11}	
0.10	k_{12}	− 0.595	− 0.529	− 0.464	− 0.398	− 0.332	− 0.582	− 0.516	− 0.450	− 0.384	− 0.318	k_{12}	0.10
	k_{13}	− 0.310	− 0.181	− 0.053	0.077	0.206	− 0.264	− 0.132	0.000	0.132	0.264	k_{13}	
	k_{11}	− 0.254	− 0.320	− 0.386	− 0.453	− 0.519	− 0.287	− 0.356	− 0.425	− 0.494	− 0.563	k_{11}	
0.15	k_{12}	− 0.576	− 0.507	− 0.439	− 0.370	− 0.301	− 0.563	− 0.494	− 0.425	− 0.356	− 0.287	k_{12}	0.15
	k_{13}	− 0.322	− 0.187	− 0.053	0.083	0.218	− 0.276	− 0.138	0.000	0.138	0.276	l_{13}	
	k_{11}	− 0.223	− 0.292	− 0.361	− 0.431	− 0.500	− 0.256	− 0.328	− 0.400	− 0.472	− 0.544	k_{11}	
0.20	k_{12}	− 0.557	− 0.485	− 0.414	− 0.342	− 0.270	− 0.544	− 0.472	− 0.400	− 0.328	− 0.256	k_{12}	0.20
	k_{13}	− 0.334	− 0.193	− 0.053	0.089	0.230	− 0.288	− 0.144	0.000	0.144	0.288	k_{13}	
	k_{11}	− 0.192	− 0.264	− 0.336	− 0.409	− 0.481	− 0.225	− 0.300	− 0.375	− 0.450	− 0.525	k_{11}	
0.25	k_{12}	− 0.536	− 0.463	− 0.389	− 0.314	− 0.239	− 0.525	− 0.450	− 0.375	− 0.300	− 0.225	k_{12}	0.25
	k_{13}	− 0.346	− 0.193	− 0.053	0.095	0.242	− 0.300	− 0.150	0.000	0.150	0.300	k_{13}	
	k_{11}	− 0.161	− 0.236	− 0.311	− 0.387	− 0.462	− 0.194	− 0.272	− 0.350	− 0.428	− 0.506	k_{11}	
0.30	k_{12}	− 0.519	− 0.441	− 0.364	− 0.283	− 0.208	− 0.506	− 0.428	− 0.350	− 0.272	− 0.194	k_{12}	0.30
	k_{13}	− 0.358	− 0.205	− 0.053	0.101	0.254	− 0.312	− 0.156	0.000	0.156	0.312	k_{13}	
	k_{11}	− 0.130	− 0.208	− 0.286	− 0.365	− 0.443	− 0.163	− 0.244	− 0.325	− 0.406	− 0.487	k_{11}	
0.35	k_{12}	− 0.500	− 0.419	− 0.339	− 0.258	− 0.177	− 0.487	− 0.406	− 0.325	− 0.244	− 0.163	k_{12}	0.35
	k_{13}	− 0.370	− 0.211	− 0.053	0.107	0.266	− 0.324	− 0.162	0.000	0.162	0.324	k_{13}	
	k_{11}	− 0.099	− 0.180	− 0.261	− 0.343	− 0.424	− 0.132	− 0.216	− 0.300	− 0.384	− 0.468	k_{11}	
0.40	k_{12}	− 0.481	− 0.397	− 0.314	− 0.230	− 0.146	− 0.468	− 0.384	− 0.300	− 0.216	− 0.132	k_{12}	0.40
	k_{13}	− 0.382	− 0.217	− 0.053	0.113	0.278	− 0.336	− 0.168	0.000	0.168	0.336	k_{13}	
	k_{11}	− 0.068	− 0.152	− 0.236	− 0.321	− 0.405	− 0.101	− 0.188	− 0.275	− 0.362	− 0.449	k_{11}	
0.45	k_{12}	− 0.462	− 0.375	− 0.289	− 0.202	− 0.115	− 0.449	− 0.362	− 0.275	− 0.188	− 0.101	k_{12}	0.45
	k_{13}	− 0.394	− 0.223	− 0.053	0.119	0.290	− 0.348	− 0.174	0.000	0.174	0.348	k_{13}	
	k_{11}	− 0.037	− 0.124	− 0.211	− 0.299	− 0.386	− 0.070	− 0.160	− 0.250	− 0.340	− 0.430	k_{11}	
0.50	k_{12}	− 0.443	− 0.353	− 0.264	− 0.174	− 0.084	− 0.430	− 0.340	− 0.250	− 0.160	− 0.070	k_{12}	0.50
	k_{13}	− 0.406	− 0.229	− 0.053	0.125	0.302	− 0.360	− 0.180	0.000	0.180	0.360	k_{13}	
	k_{11}	− 0.006	− 0.096	− 0.186	− 0.277	− 0.367	− 0.039	− 0.132	− 0.225	− 0.318	− 0.411	k_{11}	
0.55	k_{12}	− 0.424	− 0.331	− 0.239	− 0.146	− 0.053	− 0.411	− 0.318	− 0.225	− 0.132	− 0.039	k_{12}	0.55
	k_{13}	− 0.418	− 0.235	− 0.053	0.131	0.314	− 0.372	− 0.186	0.000	0.186	0.372	k_{13}	
	k_{11}	0.025	− 0.068	− 0.161	− 0.255	− 0.348	− 0.008	− 0.104	− 0.200	− 0.296	− 0.392	k_{11}	
0.60	k_{12}	− 0.405	− 0.309	− 0.214	− 0.118	− 0.022	− 0.392	− 0.296	− 0.200	− 0.104	− 0.008	k_{12}	0.60
	k_{13}	− 0.430	− 0.241	− 0.053	0.137	0.326	− 0.384	− 0.192	0.000	0.192	0.384	k_{13}	
	k_{11}	0.056	− 0.040	− 0.136	− 0.233	− 0.329	0.023	− 0.076	− 0.175	− 0.274	− 0.373	k_{11}	
0.65	k_{12}	− 0.386	− 0.287	− 0.189	− 0.090	0.009	− 0.373	− 0.274	− 0.175	− 0.076	0.023	k_{12}	0.65
	k_{13}	− 0.442	− 0.247	− 0.053	0.143	0.338	− 0.396	− 0.198	0.000	0.198	0.396	k_{13}	

To be continued Table V-5e on p. 108

Table V-5f   – 108 –

Tabelle V-5e (Continuation)

m	$r = 0.95$					$r = 1.00$					m
	$n=0.40$	$n=0.45$	$n=0.50$	$n=0.55$	$n=0.60$	$n=0.40$	$n=0.45$	$n=0.50$	$n=0.55$	$n=0.60$	
k_{11}	0.087	−0.012	−0.111	−0.211	−0.310	0.054	−0.048	−0.150	−0.252	−0.354	k_{11}
0.70 k_{12}	−0.367	−0.265	−0.164	−0.062	0.040	−0.354	−0.252	−0.150	−0.048	0.054	k_{12} 0.70
k_{13}	−0.454	−0.253	−0.053	0.149	0.350	−0.408	−0.204	0.000	0.204	0.408	k_{13}
k_{11}	0.118	0.016	−0.086	−0.189	−0.291	0.085	−0.020	−0.125	−0.230	−0.335	k_{11}
0.75 k_{12}	−0.348	−0.243	−0.139	−0.034	0.071	−0.335	−0.230	−0.125	−0.020	0.085	k_{12} 0.75
k_{13}	−0.466	−0.259	−0.053	0.155	0.362	−0.420	−0.210	0.000	0.210	0.420	k_{13}
k_{11}	0.149	0.044	−0.061	−0.167	−0.272	−0.116	0.008	−0.100	−0.208	−0.316	k_{11}
0.80 k_{12}	−0.329	−0.221	−0.114	−0.006	0.102	−0.316	−0.208	−0.100	0.008	0.116	k_{12} 0.80
k_{13}	−0.478	−0.265	−0.053	0.161	0.374	−0.432	−0.216	0.000	0.216	0.432	k_{13}
k_{11}	0.180	0.072	−0.036	−0.145	−0.253	0.147	0.036	−0.075	−0.186	−0.297	k_{11}
0.85 k_{12}	−0.310	−0.199	−0.089	0.022	0.133	−0.297	−0.186	−0.075	0.036	0.147	k_{12} 0.85
k_{13}	−0.490	−0.271	−0.053	0.167	0.386	−0.444	−0.222	0.000	0.222	0.444	k_{13}
k_{11}	0.211	0.100	−0.011	−0.123	−0.234	0.178	0.064	−0.050	−0.164	−0.278	k_{11}
0.90 k_{12}	−0.291	−0.177	−0.064	0.050	0.164	−0.278	−0.164	−0.050	0.064	0.178	k_{12} 0.90
k_{13}	−0.502	−0.277	−0.053	0.173	0.398	−0.456	−0.228	0.000	0.228	0.456	k_{13}
k_{11}	0.242	0.128	0.014	−0.101	−0.215	0.209	0.092	−0.025	−0.142	−0.259	k_{11}
0.95 k_{12}	−0.271	−0.155	−0.039	0.078	0.195	−0.259	−0.142	−0.025	0.092	0.209	k_{12} 0.95
k_{13}	−0.514	−0.283	−0.053	0.179	0.41	−0.468	−0.234	0.000	0.234	0.468	k_{13}
k_{11}	0.273	0.156	0.039	−0.079	−0.196	0.240	0.120	0.000	−0.120	−0.240	k_{11}
1.00 k_{12}	−0.253	−0.133	−0.014	0.106	0.226	−0.240	−0.120	0.000	0.120	0.240	k_{12} 1.00
k_{13}	−0.526	−0.289	−0.053	0.185	0.422	−0.480	−0.240	0.000	0.240	0.480	k_{13}

Table V-5f

Top number $= k_{14}$;   Bottom number $= k_{15}$

m	$r = 0.55$					$r = 0.60$					m
	$n=0.40$	$n=0.45$	$n=0.50$	$n=0.55$	$n=0.60$	$n=0.40$	$n=0.45$	$n=0.50$	$n=0.55$	$n=0.60$	
−0.10 k_{14}	0.178	0.178	0.179	0.181	0.182	0.187	0.188	0.190	0.192	0.195	k_{14} −0.10
k_{15}	0.282	0.270	0.258	0.246	0.233	0.283	0.272	0.260	0.248	0.235	k_{15}
−0.05 k_{14}	0.164	0.165	0.167	0.169	0.171	0.173	0.175	0.178	0.180	0.183	k_{14} −0.05
k_{15}	0.271	0.258	0.246	0.233	0.219	0.272	0.260	0.248	0.235	0.222	k_{15}
0.0 k_{14}	0.151	0.152	0.154	0.157	0.159	0.160	0.162	0.165	0.168	0.172	k_{14} 0.0
k_{15}	0.259	0.246	0.233	0.220	0.206	0.260	0.248	0.235	0.222	0.208	k_{15}
0.05 k_{14}	0.137	0.139	0.142	0.145	0.148	0.146	0.149	0.153	0.156	0.160	k_{14} 0.05
k_{15}	0.248	0.234	0.221	0.207	0.192	0.249	0.236	0.223	0.209	0.195	k_{15}
0.10 k_{14}	0.124	0.126	0.129	0.133	0.136	0.133	0.136	0.140	0.144	0.149	k_{14} 0.10
k_{15}	0.236	0.222	0.208	0.194	0.179	0.237	0.224	0.210	0.196	0.181	k_{15}
0.15 k_{14}	0.110	0.113	0.117	0.121	0.125	0.119	0.123	0.128	0.132	0.137	k_{14} 0.15
k_{15}	0.225	0.210	0.196	0.181	0.165	0.226	0.212	0.198	0.183	0.168	k_{15}
0.20 k_{14}	0.097	0.100	0.104	0.109	0.113	0.106	0.110	0.115	0.120	0.126	k_{14} 0.20
k_{15}	0.213	0.198	0.183	0.168	0.152	0.214	0.200	0.185	0.170	0.154	k_{15}
0.25 k_{14}	0.083	0.087	0.092	0.097	0.102	0.092	0.097	0.103	0.108	0.114	k_{14} 0.25
k_{15}	0.202	0.186	0.171	0.155	0.138	0.203	0.188	0.173	0.157	0.141	k_{15}
0.30 k_{14}	0.070	0.074	0.079	0.085	0.090	0.079	0.084	0.090	0.096	0.103	k_{14} 0.30
k_{15}	0.190	0.174	0.158	0.142	0.125	0.191	0.176	0.160	0.144	0.127	k_{15}

Table V-5f (Continuation)

m		r = 0.55					r = 0.60				m	
		n = 0.40	n = 0.45	n = 0.50	n = 0.55	n = 0.60	n = 0.40	n = 0.45	n = 0.50	n = 0.55	n = 0.60	
0.35	k_{14}	0.056	0.061	0.067	0.073	0.079	0.065	0.071	0.078	0.084	0.091	k_{14} 0.35
	k_{15}	0.179	0.162	0.146	0.129	0.111	0.180	0.164	0.148	0.131	0.114	k_{15}
0.40	k_{14}	0.043	0.048	0.054	0.061	0.067	0.052	0.058	0.065	0.072	0.080	k_{14} 0.40
	k_{15}	0.167	0.150	0.133	0.116	0.098	0.168	0.152	0.135	0.118	0.100	k_{15}
0.45	k_{14}	0.029	0.035	0.042	0.049	0.056	0.038	0.045	0.053	0.060	0.068	k_{14} 0.45
	k_{15}	0.156	0.138	0.121	0.103	0.084	0.157	0.140	0.123	0.105	0.087	k_{15}
0.50	k_{14}	0.016	0.022	0.029	0.037	0.044	0.025	0.032	0.040	0.482	0.057	k_{14} 0.50
	k_{15}	0.144	0.126	0.108	0.090	0.071	0.145	0.128	0.110	0.092	0.073	k_{15}
0.55	k_{14}	0.002	0.009	0.017	0.025	0.033	0.011	0.019	0.028	0.036	0.045	k_{14} 0.55
	k_{15}	0.133	0.114	0.096	0.077	0.057	0.134	0.116	0.098	0.079	0.060	k_{15}
0.60	k_{14}	− 0.011	− 0.004	0.004	0.013	0.021	− 0.002	0.006	0.015	0.024	0.034	k_{14} 0.60
	k_{15}	0.121	0.102	0.083	0.064	0.044	0.122	0.104	0.085	0.066	0.046	k_{15}
0.65	k_{14}	− 0.025	− 0.017	− 0.008	0.001	0.010	− 0.016	− 0.007	0.003	0.012	0.022	k_{14} 0.65
	k_{15}	0.110	0.090	0.071	0.051	0.030	0.111	0.092	0.073	0.053	0.033	k_{15}
0.70	k_{14}	− 0.038	− 0.030	− 0.021	− 0.011	− 0.002	− 0.029	− 0.020	− 0.010	0.000	0.011	k_{14} 0.70
	k_{15}	0.098	0.078	0.058	0.038	0.017	0.099	0.080	0.060	0.040	0.019	k_{15}
0.75	k_{14}	− 0.052	− 0.043	− 0.033	− 0.023	− 0.013	− 0.043	− 0.033	− 0.023	− 0.012	− 0.001	k_{14} 0.75
	k_{15}	0.087	0.066	0.046	0.025	0.003	0.088	0.068	0.048	0.027	0.006	k_{15}
0.80	k_{14}	− 0.065	− 0.056	− 0.046	− 0.035	− 0.025	− 0.056	− 0.046	− 0.035	− 0.024	− 0.012	k_{14} 0.80
	k_{15}	0.075	0.054	0.033	0.012	− 0.010	0.076	0.056	0.035	0.014	− 0.008	k_{15}
0.85	k_{14}	− 0.079	− 0.069	− 0.058	− 0.047	− 0.036	− 0.070	− 0.059	− 0.048	− 0.036	− 0.024	k_{14} 0.85
	k_{15}	0.064	0.042	0.021	− 0.001	− 0.024	0.065	0.044	0.023	0.001	− 0.021	k_{15}
0.90	k_{14}	− 0.092	0.082	− 0.071	− 0.059	− 0.048	− 0.083	− 0.072	− 0.060	− 0.048	− 0.035	k_{14} 0.90
	k_{15}	0.052	0.030	0.008	− 0.014	− 0.037	0.053	0.032	0.010	− 0.012	− 0.035	k_{15}
0.95	k_{14}	− 0.106	− 0.095	− 0.083	− 0.071	− 0.059	− 0.097	− 0.085	− 0.073	− 0.060	− 0.047	k_{14}
	k_{15}	0.041	0.018	− 0.004	− 0.027	− 0.051	0.042	0.020	− 0.003	− 0.025	− 0.048	k_{15} 0.95
1.00	k_{14}	− 0.119	− 0.108	− 0.096	− 0.083	− 0.071	− 0.110	− 0.098	− 0.085	− 0.072	− 0.058	k_{14} 1.00
	k_{15}	0.029	0.006	− 0.017	− 0.040	− 0.064	0.030	0.008	− 0.015	− 0.038	− 0.062	k_{15}

Table V-5g

Top number = k_{14};   Bottom number = k_{15}

m		r = 0.65					r = 0.70				m	
		n = 0.40	n = 0.45	n = 0.50	n = 0.55	n = 0.60	n = 0.40	n = 0.45	n = 0.50	n = 0.55	n = 0.60	
− 0.10	k_{14}	0.195	0.198	0.201	0.204	0.207	0.204	0.208	0.211	0.215	0.219	k_{14} − 0.10
	k_{15}	0.285	0.273	0.262	0.250	0.238	0.286	0.275	0.264	0.252	0.241	k_{15}
0.05	k_{14}	0.182	0.185	0.188	0.192	0.195	0.191	0.195	0.199	0.203	0.208	k_{14} − 0.05
	k_{15}	0.273	0.261	0.249	0.237	0.225	0.274	0.263	0.251	0.239	0.227	k_{15}
0.00	k_{14}	0.168	0.172	0.176	0.180	0.184	0.177	0.182	0.186	0.191	0.196	k_{14} 0.00
	k_{15}	0.262	0.249	0.237	0.224	0.211	0.263	0.251	0.239	0.226	0.214	k_{15}
0.05	k_{14}	0.155	0.159	0.163	0.168	0.172	0.164	0.169	0.174	0.179	0.185	k_{14} 0.05
	k_{15}	0.250	0.237	0.224	0.211	0.198	0.251	0.239	0.226	0.213	0.200	k_{15}
0.10	k_{14}	0.141	0.046	0.151	0.156	0.161	0.150	0.156	0.161	0.167	0.173	k_{14} 0.10
	k_{15}	0.239	0.225	0.212	0.198	0.184	0.240	0.227	0.214	0.200	0.187	k_{15}

8*

To be continued Table V-5g on p. 110

Table V-5g   – 110 –

Table V-5g (Continuation)

m	$r = 0.65$					$r = 0.70$						m
	$n = 0.40$	$n = 0.45$	$n = 0.50$	$n = 0.55$	$n = 0.60$	$n = 0.40$	$n = 0.45$	$n = 0.50$	$n = 0.55$	$n = 0.60$		
$0.15\ k_{14}$	0.128	0.133	0.138	0.144	0.149	0.137	0.143	0.149	0.155	0.162	k_{14}	0.15
k_{15}	0.227	0.213	0.199	0.185	0.171	0.228	0.215	0.201	0.187	0.173	k_{15}	
$0.20\ k_{14}$	0.114	0.120	0.126	0.132	0.138	0.123	0.130	0.136	0.143	0.150	k_{14}	0.20
k_{15}	0.216	0.201	0.187	0.172	0.157	0.217	0.203	0.189	0.174	0.160	k_{15}	
$0.25\ k_{14}$	0.101	0.107	0.113	0.120	0.126	0.110	0.117	0.124	0.131	0.139	k_{14}	0.25
k_{15}	0.204	0.189	0.174	0.159	0.144	0.205	0.191	0.176	0.161	0.146	k_{15}	
$0.30\ k_{14}$	0.087	0.094	0.101	0.108	0.115	0.096	0.104	0.111	0.119	0.127	k_{14}	0.30
k_{15}	0.193	0.177	0.162	0.146	0.130	0.194	0.179	0.164	0.148	0.133	k_{15}	
$0.35\ k_{14}$	0.074	0.081	0.088	0.096	0.103	0.083	0.091	0.099	0.107	0.116	k_{14}	0.35
k_{15}	0.181	0.165	0.149	0.133	0.117	0.182	0.167	0.151	0.135	0.119	k_{15}	
$0.40\ k_{14}$	0.060	0.068	0.076	0.084	0.092	0.069	0.078	0.086	0.095	0.104	k_{14}	0.40
k_{15}	0.170	0.153	0.137	0.120	0.103	0.171	0.155	0.139	0.122	0.106	k_{15}	
$0.45\ k_{14}$	0.047	0.055	0.063	0.072	0.080	0.056	0.065	0.074	0.083	0.093	k_{14}	0.45
k_{15}	0.158	0.141	0.124	0.107	0.090	0.159	0.143	0.126	0.109	0.092	k_{15}	
$0.50\ k_{14}$	0.033	0.042	0.051	0.060	0.069	0.042	0.052	0.061	0.071	0.081	k_{14}	0.50
k_{15}	0.147	0.129	0.112	0.094	0.076	0.148	0.131	0.114	0.096	0.079	k_{15}	
$0.55\ k_{14}$	0.020	0.029	0.038	0.048	0.057	0.029	0.039	0.049	0.059	0.070	k_{14}	0.55
k_{15}	0.135	0.117	0.099	0.081	0.063	0.136	0.119	0.101	0.083	0.065	k_{15}	
$0.60\ k_{14}$	0.006	0.016	0.026	0.036	0.046	0.015	0.026	0.036	0.047	0.058	k_{14}	0.60
k_{15}	0.124	0.105	0.087	0.068	0.049	0.125	0.107	0.089	0.070	0.052	k_{15}	
$0.65\ k_{14}$	– 0.007	0.003	0.013	0.024	0.034	0.002	0.013	0.024	0.035	0.047	k_{14}	0.65
k_{15}	0.112	0.093	0.074	0.055	0.036	0.113	0.095	0.076	0.057	0.038	k_{15}	
$0.70\ k_{14}$	– 0.021	– 0.010	0.001	0.012	0.023	– 0.012	0.000	0.011	0.023	0.035	k_{14}	0.70
k_{15}	0.101	0.081	0.062	0.042	0.022	0.102	0.083	0.064	0.044	0.025	k_{15}	
$0.75\ k_{14}$	– 0.034	– 0.023	– 0.012	– 0.000	0.011	– 0.025	– 0.013	– 0.001	0.011	0.024	k_{14}	0.75
k_{15}	0.089	0.069	0.049	0.029	0.009	0.090	0.071	0.051	0.031	0.011	k_{15}	
$0.80\ k_{14}$	– 0.048	– 0.036	– 0.024	– 0.012	0.000	– 0.039	– 0.026	– 0.014	– 0.001	0.012	k_{14}	0.80
k_{15}	0.078	0.057	0.037	0.016	– 0.005	0.079	0.059	0.039	0.018	– 0.002	k_{15}	
$0.85\ k_{14}$	– 0.061	– 0.049	– 0.037	– 0.024	– 0.012	– 0.052	– 0.039	– 0.026	– 0.013	0.001	k_{14}	0.85
k_{15}	0.066	0.045	0.024	0.003	– 0.018	0.067	0.047	0.026	0.005	– 0.016	k_{15}	
$0.90\ k_{14}$	– 0.075	– 0.062	– 0.049	– 0.036	– 0.023	– 0.066	– 0.052	– 0.039	– 0.025	– 0.011	k_{14}	0.90
k_{15}	0.055	0.033	0.012	– 0.010	– 0.032	0.056	0.035	0.014	– 0.008	– 0.029	k_{15}	
$0.95\ k_{14}$	– 0.089	– 0.075	– 0.062	– 0.048	– 0.035	– 0.079	– 0.065	– 0.051	– 0.037	– 0.022	k_{14}	0.95
k_{15}	0.043	0.021	– 0.001	– 0.023	– 0.045	0.044	0.023	0.001	– 0.021	– 0.043	k_{15}	
$1.00\ k_{14}$	– 0.102	– 0.088	– 0.074	– 0.060	– 0.046	– 0.093	– 0.078	– 0.064	– 0.049	– 0.034	k_{14}	1.00
k_{15}	0.032	0.009	– 0.013	– 0.036	– 0.059	0.033	0.011	– 0.011	– 0.034	– 0.056	k_{15}	

Table V-5h

Top number $= k_{14}$;   Bottom number $= k_{15}$

m		r = 0.75					r = 0.80					m
		n = 0.40	n = 0.45	n = 0.50	n = 0.55	n = 0.60	n = 0.40	n = 0.45	n = 0.50	n = 0.55	n = 0.60	
-0.10	k_{14}	0.213	0.217	0.222	0.227	0.232	0.222	0.227	0.233	0.238	0.244	k_{14} 0.10
	k_{15}	0.287	0.276	0.266	0.255	0.244	0.288	0.278	0.268	0.257	0.246	k_{15}
-0.05	k_{14}	0.200	0.204	0.209	0.215	0.220	0.208	0.214	0.220	0.226	0.232	k_{14} -0.05
	k_{15}	0.276	0.264	0.253	0.242	0.230	0.277	0.266	0.255	0.244	0.233	k_{15}
0.00	k_{14}	0.186	0.191	0.197	0.203	0.209	0.195	0.201	0.208	0.214	0.221	k_{14} 0.00
	k_{15}	0.264	0.252	0.241	0.229	0.217	0.265	0.254	0.243	0.231	0.219	k_{15}
0.05	k_{14}	0.173	0.178	0.184	0.191	0.197	0.181	0.188	0.195	0.202	0.209	k_{14} 0.05
	k_{15}	0.253	0.240	0.228	0.216	0.203	0.254	0.242	0.230	0.218	0.206	k_{15}
0.10	k_{14}	0.159	0.165	0.172	0.179	0.186	0.168	0.175	0.183	0.190	0.198	k_{14} 0.10
	k_{15}	0.241	0.228	0.216	0.203	0.190	0.242	0.230	0.218	0.205	0.192	k_{15}
0.15	k_{14}	0.146	0.152	0.159	0.167	0.174	0.154	0.162	0.170	0.178	0.186	k_{14} 0.15
	k_{15}	0.230	0.216	0.203	0.190	0.176	0.231	0.218	0.205	0.192	0.179	k_{15}
0.20	k_{14}	0.132	0.139	0.147	0.155	0.163	0.141	0.149	0.158	0.166	0.175	k_{14} 0.20
	k_{15}	0.218	0.204	0.191	0.177	0.163	0.219	0.206	0.193	0.179	0.165	k_{15}
0.25	k_{14}	0.119	0.126	0.134	0.143	0.151	0.127	0.136	0.145	0.154	0.163	k_{14} 0.25
	k_{15}	0.207	0.192	0.178	0.164	0.149	0.208	0.194	0.180	0.166	0.152	k_{15}
0.30	k_{14}	0.105	0.113	0.122	0.131	0.140	0.114	0.123	0.133	0.142	0.152	k_{14} 0.30
	k_{15}	0.195	0.180	0.166	0.151	0.136	0.196	0.182	0.168	0.153	0.138	k_{15}
0.35	k_{14}	0.092	0.100	0.109	0.119	0.128	0.100	0.110	0.120	0.130	0.140	k_{14} 0.35
	k_{15}	0.184	0.168	0.153	0.138	0.122	0.185	0.170	0.155	0.140	0.125	k_{15}
0.40	k_{14}	0.078	0.087	0.097	0.107	0.117	0.087	0.097	0.108	0.118	0.129	k_{14} 0.40
	k_{15}	0.172	0.156	0.141	0.125	0.109	0.173	0.158	0.143	0.127	0.111	k_{15}
0.45	k_{14}	0.065	0.074	0.084	0.095	0.105	0.073	0.084	0.095	0.106	0.117	k_{14} 0.45
	k_{15}	0.161	0.144	0.128	0.112	0.095	0.162	0.146	0.130	0.114	0.098	k_{15}
0.50	k_{14}	0.051	0.061	0.072	0.083	0.094	0.060	0.071	0.083	0.094	0.106	k_{14} 0.50
	k_{15}	0.149	0.132	0.116	0.099	0.082	0.150	0.134	0.118	0.101	0.084	k_{15}
0.55	k_{14}	0.038	0.048	0.059	0.071	0.082	0.046	0.058	0.070	0.082	0.094	k_{14} 0.55
	k_{15}	0.138	0.120	0.103	0.086	0.068	0.139	0.122	0.105	0.088	0.071	k_{15}
0.60	k_{14}	0.024	0.035	0.047	0.059	0.071	0.033	0.045	0.058	0.070	0.083	k_{14} 0.60
	k_{15}	0.126	0.108	0.091	0.073	0.055	0.127	0.110	0.093	0.075	0.057	k_{15}
0.65	k_{14}	0.011	0.022	0.034	0.047	0.059	0.019	0.032	0.045	0.058	0.071	k_{14} 0.65
	k_{15}	0.115	0.096	0.078	0.060	0.041	0.116	0.098	0.080	0.062	0.044	k_{15}
0.70	k_{14}	-0.003	0.009	0.022	0.035	0.048	0.006	0.019	0.033	0.046	0.060	k_{14} 0.70
	k_{15}	0.103	0.084	0.066	0.047	0.028	0.104	0.086	0.068	0.049	0.030	k_{15}
0.75	k_{14}	-0.017	-0.004	0.009	0.023	0.036	-0.008	0.006	0.020	0.034	0.048	k_{14} 0.75
	k_{15}	0.092	0.072	0.053	0.034	0.014	0.093	0.074	0.055	0.036	0.017	k_{15}
0.80	k_{14}	-0.030	-0.017	-0.003	0.011	0.025	-0.021	-0.007	0.008	0.022	0.037	k_{14} 0.80
	k_{15}	0.080	0.060	0.041	0.021	0.001	0.081	0.062	0.043	0.023	0.003	k_{15}
0.85	k_{14}	-0.044	-0.030	-0.016	-0.001	0.013	-0.035	-0.020	-0.005	0.010	0.025	k_{14} 0.85
	k_{15}	0.069	0.048	0.028	0.008	-0.013	0.070	0.050	0.030	0.010	-0.010	k_{15}
0.90	k_{14}	-0.057	-0.043	-0.028	-0.013	0.002	-0.048	-0.033	-0.018	-0.002	0.014	k_{14} 0.90
	k_{15}	0.057	0.036	0.016	-0.005	-0.027	0.058	0.038	0.018	-0.003	-0.024	k_{15}
0.95	k_{14}	-0.071	0.056	0.041	-0.025	-0.010	-0.062	-0.046	0.030	-0.014	0.002	k_{14} 0.95
	k_{15}	0.046	0.024	0.003	-0.018	-0.040	0.047	0.026	0.005	-0.016	-0.037	k_{15}
1.00	k_{14}	-0.084	-0.069	-0.053	-0.037	-0.022	-0.075	-0.059	-0.043	-0.026	-0.009	k_{14} 1.00
	k_{15}	0.034	0.012	-0.009	-0.031	-0.054	0.035	0.014	-0.008	-0.029	-0.051	k_{15}

Table V-5i — 112 —

Table V-5i

Top number $= k_{14}$;   Bottom number $= k_{15}$

m		$r = 0.85$					$r = 0.90$				m
		$n=0.40$	$n=0.45$	$n=0.50$	$n=0.55$	$n=0.60$	$n=0.40$	$n=0.45$	$n=0.50$	$n=0.55$	$n=0.60$	
-0.10	k_{14}	0.231	0.236	0.243	0.250	0.256	0.239	0.247	0.254	0.261	0.268	k_{14} -0.10
	k_{15}	0.289	0.279	0.269	0.259	0.249	0.291	0.281	0.271	0.262	0.252	k_{15}
0.05	k_{14}	0.217	0.224	0.231	0.238	0.245	0.226	0.234	0.241	0.249	0.257	k_{14} -0.05
	k_{15}	0.278	0.267	0.257	0.246	0.235	0.279	0.269	0.259	0.249	0.238	k_{15}
0.00	k_{14}	0.204	0.211	0.218	0.226	0.233	0.212	0.221	0.229	0.237	0.245	k_{14} 0.00
	k_{15}	0.266	0.255	0.244	0.233	0.222	0.268	0.257	0.246	0.236	0.225	k_{15}
0.05	k_{14}	0.190	0.198	0.206	0.214	0.222	0.199	0.208	0.216	0.225	0.234	k_{14} 0.05
	k_{15}	0.255	0.243	0.232	0.220	0.208	0.256	0.245	0.234	0.223	0.211	k_{15}
0.10	k_{14}	0.177	0.185	0.193	0.202	0.210	0.185	0.195	0.204	0.213	0.222	k_{14} 0.10
	k_{15}	0.243	0.231	0.219	0.207	0.195	0.245	0.233	0.221	0.210	0.198	k_{15}
0.15	k_{14}	0.163	0.172	0.181	0.190	0.199	0.172	0.182	0.191	0.201	0.211	k_{14} 0.15
	k_{15}	0.232	0.219	0.207	0.194	0.181	0.233	0.221	0.209	0.197	0.184	k_{15}
0.20	k_{14}	0.150	0.159	0.168	0.178	0.187	0.158	0.169	0.179	0.189	0.199	k_{14} 0.20
	k_{15}	0.220	0.207	0.194	0.181	0.168	0.222	0.209	0.196	0.184	0.171	k_{15}
0.25	k_{14}	0.136	0.146	0.156	0.166	0.176	0.145	0.156	0.166	0.177	0.188	k_{14} 0.25
	k_{15}	0.209	0.195	0.182	0.168	0.154	0.210	0.197	0.184	0.171	0.157	k_{15}
0.30	k_{14}	0.123	0.133	0.143	0.154	0.164	0.131	0.143	0.154	0.165	0.176	k_{14} 0.30
	k_{15}	0.197	0.183	0.169	0.155	0.141	0.199	0.185	0.171	0.158	0.144	k_{15}
0.35	k_{14}	0.109	0.120	0.121	0.142	0.153	0.118	0.130	0.141	0.153	0.165	k_{14} 0.35
	k_{15}	0.186	0.171	0.157	0.142	0.127	0.187	0.173	0.159	0.145	0.130	k_{15}
0.40	k_{14}	0.096	0.107	0.118	0.130	0.141	0.104	0.117	0.129	0.141	0.153	k_{14} 0.40
	k_{15}	0.174	0.159	0.144	0.129	0.114	0.176	0.161	0.146	0.132	0.117	k_{15}
0.45	k_{14}	0.082	0.094	0.106	0.118	0.130	0.091	0.104	0.116	0.129	0.142	k_{14} 0.45
	k_{15}	0.163	0.147	0.132	0.116	0.100	0.164	0.149	0.134	0.119	0.103	k_{15}
0.50	k_{14}	0.069	0.081	0.093	0.106	0.118	0.077	0.091	0.104	0.117	0.130	k_{14} 0.50
	k_{15}	0.151	0.135	0.119	0.103	0.087	0.153	0.137	0.121	0.106	0.090	k_{15}
0.55	k_{14}	0.055	0.068	0.081	0.094	0.107	0.064	0.078	0.091	0.105	0.119	k_{14} 0.55
	k_{15}	0.140	0.123	0.107	0.090	0.073	0.141	0.125	0.109	0.093	0.076	k_{15}
0.60	k_{14}	0.042	0.055	0.068	0.082	0.095	0.050	0.065	0.079	0.093	0.107	k_{14} 0.60
	k_{15}	0.128	0.111	0.094	0.077	0.060	0.130	0.113	0.096	0.080	0.063	
0.65	k_{14}	0.028	0.042	0.056	0.070	0.084	0.037	0.052	0.066	0.081	0.096	k_{14} 0.65
	k_{15}	0.117	0.099	0.082	0.064	0.046	0.118	0.101	0.084	0.067	0.049	k_{15}
0.70	k_{14}	0.015	0.029	0.043	0.058	0.072	0.023	0.039	0.054	0.069	0.084	k_{14} 0.70
	k_{15}	0.105	0.087	0.069	0.051	0.033	0.107	0.089	0.071	0.054	0.036	k_{15}
0.75	k_{14}	0.001	0.016	0.031	0.046	0.061	0.010	0.026	0.041	0.057	0.073	k_{14} 0.75
	k_{15}	0.094	0.075	0.057	0.038	0.019	0.095	0.077	0.059	0.041	0.022	k_{15}
0.80	k_{14}	-0.012	0.003	0.018	0.034	0.049	-0.004	0.013	0.029	0.045	0.061	k_{14} 0.80
	k_{15}	0.082	0.063	0.044	0.025	0.006	0.084	0.065	0.046	0.028	0.009	k_{15}
0.85	k_{14}	-0.026	-0.010	0.006	0.022	0.038	-0.017	-0.001	0.016	0.033	0.050	k_{14} 0.85
	k_{15}	0.071	0.051	0.032	0.012	-0.008	0.072	0.053	0.034	0.015	-0.005	k_{15}
0.90	k_{14}	-0.039	-0.023	-0.007	0.010	0.026	-0.031	-0.014	0.004	0.021	0.038	k_{14} 0.90
	k_{15}	0.059	0.039	0.019	-0.001	-0.021	0.061	0.041	0.021	0.002	-0.018	k_{15}
0.95	k_{14}	-0.053	-0.036	-0.019	-0.002	0.015	-0.044	-0.027	-0.009	0.009	0.027	k_{14} 0.95
	k_{15}	0.048	0.027	0.007	-0.014	-0.035	0.049	0.029	0.009	-0.012	-0.032	k_{15}
1.00	k_{14}	-0.066	-0.049	-0.032	-0.014	0.003	-0.058	-0.040	-0.021	-0.003	0.015	k_{14} 1.00
	k_{15}	0.036	0.015	-0.006	-0.027	-0.048	0.038	0.017	-0.004	-0.025	-0.045	k_{15}

Table V-5k

Top number $= k_{14}$; Bottom number $= k_{14}$

m	$r = 0.95$					$r = 1.00$					m
	$n=0.40$	$n=0.45$	$n=0.50$	$n=0.55$	$n=0.60$	$n=0.40$	$n=0.45$	$n=0.50$	$n=0.55$	$n=0.60$	
$-0.10\ k_{14}$	0.248	0.256	0.264	0.273	0.281	0.257	0.266	0.275	0.284	0.293	$k_{14}\ -0.10$
k_{15}	0.292	0.283	0.273	0.264	0.254	0.293	0.284	0.275	0.266	0.257	k_{15}
$-0.05\ k_{14}$	0.235	0.243	0.252	0.261	0.269	0.244	0.253	0.263	0.272	0.282	$k_{14}\ -0.05$
k_{15}	0.280	0.271	0.261	0.251	0.241	0.282	0.272	0.263	0.253	0.244	k_{15}
$0.00\ k_{14}$	0.221	0.230	0.239	0.249	0.258	0.230	0.240	0.250	0.260	0.270	$k_{14}\ 0.00$
k_{15}	0.269	0.259	0.248	0.238	0.227	0.270	0.260	0.250	0.240	0.230	k_{15}
$0.05\ k_{14}$	0.208	0.217	0.227	0.237	0.246	0.217	0.227	0.238	0.248	0.259	$k_{14}\ 0.05$
k_{15}	0.257	0.247	0.236	0.225	0.214	0.259	0.248	0.238	0.227	0.217	k_{15}
$0.10\ k_{14}$	0.194	0.204	0.214	0.225	0.235	0.203	0.214	0.225	0.236	0.247	$k_{14}\ 0.10$
k_{15}	0.246	0.235	0.223	0.212	0.200	0.247	0.236	0.225	0.214	0.203	k_{15}
$0.15\ k_{14}$	0.181	0.191	0.202	0.213	0.223	0.190	0.201	0.213	0.224	0.236	$k_{14}\ 0.15$
k_{15}	0.234	0.223	0.211	0.199	0.187	0.236	0.224	0.213	0.201	0.190	k_{15}
$0.20\ k_{14}$	0.167	0.178	0.189	0.201	0.212	0.176	0.188	0.200	0.212	0.224	$k_{14}\ 0.20$
k_{15}	0.223	0.211	0.198	0.186	0.173	0.224	0.212	0.200	0.188	0.176	k_{15}
$0.25\ k_{14}$	0.154	0.165	0.177	0.189	0.200	0.163	0.175	0.188	0.200	0.213	$k_{14}\ 0.25$
k_{15}	0.211	0.199	0.186	0.173	0.160	0.213	0.200	0.188	0.175	0.163	k_{15}
$0.30\ k_{14}$	0.140	0.152	0.164	0.177	0.189	0.149	0.162	0.175	0.188	0.201	$k_{14}\ 0.30$
k_{15}	0.200	0.187	0.173	0.160	0.146	0.201	0.188	0.175	0.162	0.149	k_{15}
$0.35\ k_{14}$	0.127	0.139	0.152	0.165	0.177	0.136	0.149	0.163	0.176	0.190	$k_{14}\ 0.35$
k_{15}	0.188	0.175	0.161	0.147	0.133	0.190	0.176	0.163	0.149	0.136	k_{15}
$0.40\ k_{14}$	0.113	0.126	0.139	0.153	0.166	0.122	0.136	0.150	0.164	0.178	$k_{14}\ 0.40$
k_{15}	0.177	0.163	0.148	0.134	0.119	0.178	0.164	0.150	0.136	0.122	k_{15}
$0.45\ k_{14}$	0.100	0.113	0.127	0.141	0.154	0.109	0.123	0.138	0.152	0.167	$k_{14}\ 0.45$
k_{15}	0.165	0.151	0.136	0.121	0.106	0.167	0.152	0.138	0.123	0.109	k_{15}
$0.50\ k_{14}$	0.086	0.100	0.114	0.129	0.143	0.095	0.110	0.125	0.140	0.155	$k_{14}\ 0.50$
k_{15}	0.154	0.139	0.123	0.108	0.092	0.155	0.140	0.125	0.110	0.095	k_{15}
$0.55\ k_{14}$	0.073	0.087	0.102	0.117	0.131	0.082	0.097	0.113	0.128	0.144	$k_{14}\ 0.55$
k_{15}	0.142	0.127	0.111	0.095	0.079	0.144	0.128	0.113	0.097	0.081	k_{15}
$0.60\ k_{14}$	0.059	0.074	0.089	0.105	0.120	0.068	0.084	0.100	0.116	0.132	$k_{14}\ 0.60$
k_{15}	0.131	0.115	0.098	0.082	0.065	0.132	0.116	0.100	0.084	0.068	k_{15}
$0.65\ k_{14}$	0.046	0.061	0.077	0.093	0.108	0.055	0.071	0.088	0.104	0.121	$k_{14}\ 0.65$
k_{15}	0.119	0.103	0.086	0.069	0.052	0.121	0.104	0.088	0.071	0.055	k_{15}
$0.70\ k_{14}$	0.032	0.048	0.064	0.081	0.097	0.041	0.058	0.075	0.092	0.109	$k_{14}\ 0.70$
k_{15}	0.108	0.091	0.073	0.056	0.038	0.109	0.092	0.075	0.058	0.041	k_{15}
$0.75\ k_{14}$	0.019	0.035	0.052	0.069	0.085	0.028	0.045	0.063	0.080	0.098	$k_{14}\ 0.75$
k_{15}	0.096	0.079	0.061	0.043	0.025	0.098	0.080	0.063	0.045	0.028	k_{15}
$0.80\ k_{14}$	0.005	0.022	0.039	0.057	0.074	0.014	0.032	0.050	0.068	0.086	$k_{14}\ 0.80$
k_{15}	0.085	0.067	0.048	0.030	0.011	0.086	0.068	0.050	0.032	0.014	k_{15}
$0.85\ k_{14}$	-0.008	0.009	0.027	0.045	0.062	0.001	0.019	0.038	0.056	0.075	$k_{14}\ 0.85$
k_{15}	0.073	0.055	0.036	0.017	-0.002	0.075	0.056	0.038	0.019	0.001	k_{15}
$0.90\ k_{14}$	-0.022	-0.004	0.014	0.033	0.051	-0.013	0.006	0.025	0.044	0.063	$k_{14}\ 0.90$
k_{15}	0.062	0.043	0.023	0.004	-0.016	0.063	0.044	0.025	0.006	-0.013	k_{15}
$0.95\ k_{14}$	-0.035	-0.017	0.002	0.021	0.039	-0.027	-0.007	0.013	0.032	0.052	$k_{14}\ 0.95$
k_{15}	0.050	0.031	0.011	-0.009	-0.029	0.052	0.032	0.013	-0.007	-0.027	k_{15}
$1.00\ k_{14}$	-0.049	-0.030	-0.011	0.009	0.028	-0.040	-0.020	0.000	0.020	0.040	$k_{14}\ 1.00$
k_{15}	0.039	0.019	-0.002	-0.022	-0.043	0.040	0.020	0.000	-0.020	-0.040	k_{15}